Acquisition
of
Defense Systems

Edited by
J. S. Przemieniecki
Air Force Institute of Technology

EDUCATION SERIES
J. S. Przemieniecki
Series Editor-in-Chief
Air Force Institute of Technology
Wright-Patterson Air Force Base, Ohio

Published by
American Institute of Aeronautics and Astronautics, Inc.,
370 L'Enfant Promenade, SW, Washington, DC 20024-2518

American Institute of Aeronautics and Astronautics, Inc. Washington, DC

Library of Congress Cataloging in Publication Data

Acquisition of Defense Systems, / edited by J.S. Przemieniecki.

p.cm.–(AIAA education series) Includes bibliographical references.
1. United States. Air Force–Procurement.   2. United States. Air Force–Weapon Systems.

I. Przemieniecki, J. S.   II. Series.
UG1123.A28 1993   358.4' 16212' 0973–dc20   93-27833
ISBN 1-56347-069-1

The opinions and conclusions in this book are those of the authors and are not intended to represent the official position of the Department of Defense, United States Air Force, or any other government agency. This book is published in the Education Series of the American Institute of Aeronautics and Astronautics (AIAA) under a Cooperative Research and Development Agreement (CRADA) with the United States Air Force.

Space Vehicle Design
  Michael D. Griffin and James R. French, 1991
Inlets for Supersonic Missiles
  John J. Mahoney, 1991
Defense Analyses Software
  J. S. Przemieniecki, 1991
Critical Technologies for National Defense
  Air Force Institute of Technology, 1991
Orbital Mechanics
  Vladimir A. Chobotov, 1991
Nonlinear Analysis of Shell Structures
  Anthony N. Palazotto and Scott T. Dennis, 1992
Optimization of Observation and Control Processes
  Veniamin V. Malyshev, Mihkail N. Krasilshikov, and Valeri I. Karlov,
    1992
Aircraft Design: A Conceptual Approach
  Second Edition
  Daniel P. Raymer, 1992
Rotary Wing Structural Dynamics and Aeroelasticity
  Richard L. Bielawa, 1992
Spacecraft Mission Design
  Charles D. Brown, 1992
Introduction to Dynamics and Control of Flexible Structures
  John L. Junkins and Youdan Kim, 1993
Dynamics of Atmospheric Re-Entry
  Frank J. Regan and Satya M. Anandakrishnan, 1993
Acquisition of Defense Systems
  J. S. Przemieniecki, 1993

Published by
American Institute of Aeronautics and Astronautics, Inc., Washington, DC

FOREWORD

The publication of the *Acquisition of Defense Systems* textbook prepared by the Air Force Institute of Technology (AFIT) is a significant contribution to technical management literature for acquisition of defense systems. This text will contribute to a better understanding of the underlying management principles and regulations used for the development of major systems in the Department of Defense. As the nation moves toward a smaller defense force, the establishment of a streamlined, effective, and efficient process for acquiring future defense systems becomes even more important. This requires an investment strategy that emphasizes the most promising areas of science and technology which have the greatest potential for impact on future military capabilities. The most important consideration, however, is to ensure that the acquisition process is such that these new capabilities are secured within the allocated budget. The new initiatives under development within the Department of Defense will certainly help to achieve this goal.

The management of the acquisition process for defense systems described in this AFIT text represents the Department of Defense approach to the process based on the current laws and legislative directives of the U.S. Congress. For this reason, *Acquisition of Defense Systems* is recommended reading for Armed Forces senior managers in charge of weapon systems development and acquisition. It is also a valuable text for managers in a wide range of industries.

Jay W. Kelley
Lt. Gen., USAF
Air University Commander

PREFACE

This textbook has been prepared by the Office of the Senior Dean, Air Force Institute of Technology (AFIT), Wright-Patterson Air Force Base, Ohio, with the specific objective of providing information on the latest concepts and procedures for the acquisition of defense systems conforming with the current policies and directives of the Department of Defense. For the purpose of this text, a very broad definition of acquisition is used. It is defined here as the planning, design, development, testing, contracting, production, introduction, acquisition logistics support, deployment, disposal of systems, equipment, facilities, supplies, or services that are intended for use in, or in support of, military missions. Although many of the examples discussed in the text are taken from the Air Force, the general description of the process is also applicable to systems from other services. In developing this text, lecture notes from many different AFIT courses in acquisition, logistics, management, contracting, etc., as well as source materials provided by the Department of Defense and various Army, Navy, and Air Force organizations have been used. Without the availability of these materials this book project would have never been accomplished.

The text begins by introducing the requirements and acquisition process, and then explaining that the acquisition of defense systems has its roots in our National Security Objectives typically identified in the National Security Strategy of the United States, a policy document published by the White House. Next, the formal framework of the acquisition process is described: Concept Exploration and Definition, Demonstration/Validation, Engineering and Manufacturing Development, Production and Deployment, and Operations and Support. Beyond the formal framework, however, there are numerous activities supporting the process and allowing a product that is affordable, reliable, and technologically superior to reach its final destination in operational units. These activities are described in chapters on program management, total quality, integrated weapon system management, science and technology management, financial management, program control, contracting, systems engineering, configuration management, data management, computer hardware and software, integrated logistics support, manufacturing management, test and evaluation, maintenance and operations, environmental issues, international programs, and acquisition education. These chapters provide a comprehensive overview of all the essential processes and concepts that allow building a quality product for national defense.

The main educational value of this text is in the comprehensive description of the whole process of developing and acquiring new defense systems. This text should provide defense system managers, both in DoD and in the defense industry, with an overview of their own primary area of expertise as well as other related activities affecting the quality of the system. The text can also be used in many DoD courses, either as a primary text or as a supplemental reading.

The *Acquisition of Defense Systems* text was made possible through the dedicated support and contributions from the AFIT faculty. In carrying out the arduous task of writing this book, the following AFIT faculty members deserve special recognition: Richard A. Andrews, Lt. Col. Christopher D. Arnold, Dr. Craig M. Brandt, Patrick M. Bresnahan, Dr. Anthony P. D'Angelo, William A. Dean, Richard A. DiLorenzo, Capt. James F. Donaghue, Lt. Col. Charles M. Farr, Daniel V. Ferens, John W. Garrett, Lt. Col. Mark N. Goltz, Lt. Col. Michael E. Heberling, Capt. Brian W. Holmgren, Maj. Dennis L. Hull, Capt. Marsha Kwolek, , Lt. Col. Phillip E. Miller, Arthur A. Munguia, Dr. William C. Pursch, Lt. Col. William L. Schneider, Col. Paul T. Welch, and Dr. Ben L. Williams. Thanks are also extended to many other individuals in the Air Force and DoD who provided their valuable comments and suggestions on this project. In particular, the following should be recognized for their support: Col. Blaise J. Durante, Secretary of the Air Force Office; Dr. Donald C. Fraser, former Principal Deputy Under Secretary of Defense Acquisition; CDR Timothy J. Harp, Department of Defense; Gerald E. Keightley, Defense Acquisition University; Lt. Gen. Edwin S. Leland (U.S. Army), Joint Chiefs of Staff (J-5); Philip P. Panzarella, Air Force Materiel Command (AFMC); Dr. G. Keith Richey, Wright Laboratory (AFMC); and Maj. Gen. Robert R. Rankine, Jr., AFMC.

J. S. Przemieniecki
Institute Senior Dean and Scientific Advisor
Air Force Institute of Technology
Air University

CONTENTS

CONTENTS

Chapter 1

DEFENSE REQUIREMENTS PROCESS

1.1   Introduction

The dramatic collapse of the Soviet Union and the spread of democracy and freedom throughout the former Soviet Block has transformed the world, removing the old East-West confrontation and ending the Cold War. This change in the strategic environment gives the United States the opportunity to reduce the size of its armed forces while still retaining the military capability to protect its interests. The end of Cold War, however, has not removed all threats to security and vital interests of the U.S., as evidenced by the invasion of Kuwait in 1990. Of particular concern is the proliferation of highly sophisticated weapons, which could be used in regional crises and conflicts involving U.S. forces. This means that we must continue to proceed with modernization of our forces, reshaping our military structure and maintaining a technological edge over our adversaries, albeit at a more deliberate pace than during the Cold War era pressures to move new technology weapons quickly to production to stay ahead of Soviet modernization effort. Defense must now be planned for a new era in a rapidly changing world.[1]

1.2   National Security Policy*

The acquisition process is actually an extension of the National Security Policy process. To facilitate discussion we will draw upon the model in Fig. 1.1, adapted from one developed by Harold McCard, President of Textron Defense Systems.[3] This model focuses only on the military instrument of power. It does not address plans or actions that use the economic and political instruments of power, which are of course other possible alternatives for achieving national security objectives.[4]

---

*Based on Ref. 2.

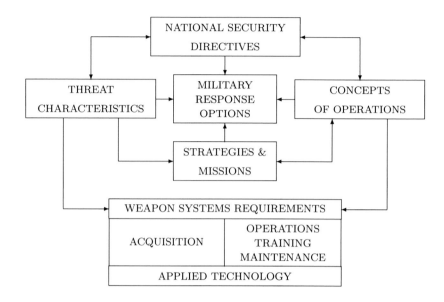

Fig. 1.1: Acquisition as an extension of the National Security Policy process.

Walking our way through the model, we see that national security objectives, once translated into directives, require some sort of military response option. The formulation of these directives are affected by two conditions, the threat and the concept of operations.

National security objectives and their subsequent directives are derived from an assessment of the threat. The purpose of these directives is to insure that our national security will be maintained. An example of the influence of the threat on our National Security Objectives can be seen in the content of the White House's National Security Strategy of the United States. The 1987 and 1988 publications identified Soviet expansion and their military capability as the principle threat to our National Security.[5,6] However, the 1990/1991 publications made no mention of the Soviet Union being a direct threat.[7,8]

The threat and our concepts of operations also affect our strategies and missions. The recent Strategic Arms Reduction Treaty (START) and conventional forces reduction treaty reflect a perceived decrease in the threat. With the reduction of strategic weapons and a likely increase in third world conflicts, we will be planning for scenarios requiring rapid deployment, massive airlift, and mobile, high firepower systems instead of a European scenario between industrially advanced nations with high technology and heavy armor.

The same situation exists between strategies and missions, and our concepts of operations. One can readily see that strategies and missions (the plan for how we will meet our objectives) drive the kinds of operations we plan for. However, the reverse is also a consideration. It does no good to develop plans that we can not execute due to a lack of forces. Resource availability sets limits on the planning of strategies and missions.[9] This planning process is governed by supplementing regulations in individual services. For example, the relevant documents for the Air Force are Air Force Policy Directive 10-6 and Air Force Instruction 10-601, which deals with the mission needs and operational requirements process.[10,11]

All of these are deciding factors, neither one singularly more important than any other in the military response options available to our leaders. The analysts' assessment of the threat operations are reflected in the weapon system we require. This should be intuitively obvious from McCard's model and the foregoing discussion. What is not readily obvious is the impact of a deficiency. What about the scenario where an assessment reveals a threat for which we have not prepared, or a concept of operation for which we do not have the necessary weapon capability?

When current capability is not enough to support response options for achieving our National Security Directives, we have the additional options of either changing our operations, training, and maintenance, modifying an existing system, or acquiring a new weapons system. The first option is usually the least costly and thus receives first consideration.

In the case of operations, if we have a bomber whose effective range is inadequate for the intended mission, we have the choice of stationing it closer to its target or modifying the aircraft with a newer and more efficient engine. In the case of training, suppose an adversary installs a new radar that will detect aircraft as low as 100 feet. One method to address this threat would be to train to fly missions at 50 to 75 feet. Another method would be to employ some type of stand-off penetrating missile. In the case of maintenance, suppose that we find the radios in our fighter/bombers achieve only half of the expected mean time between failure. An option here could be to double the number of purchased contractor repairs or we could modify the aircraft by purchasing an off the shelf, more technologically advanced radio.

If, after all of this, there is still a threat that cannot be met, acquiring a new, more technologically capable weapon system is likely to be the best way to acquire the technological superiority. This is, of course, a long term solution requiring significant resources. As the model indicates, the appli-

cation of new technology will probably undergird any alternative solutions
that might be implemented.

Since Vietnam we have been buying "smarter" and more technologically
complex weapons. The cost of these weapons has been high, prompting critics to question their usefulness. Before Desert Storm, critics were asking
if the benefits of the new weapons systems justified their enormous cost,
especially since they had never been employed in combat. Desert Storm,
however, changed this criticism. Some of our newest and most sophisticated systems were displayed, tested, and proven during Desert Storm, and
they proved the wisdom of continually updating and modifying our mature weapon systems (i.e., B-52, F- 111, AWACS, C-141, etc.). The lives
saved and success of the operation reinforced the benefits of having high
technology weapons.

1.3   Formulation of National Military Strategy

Figure 1.2 shows schematically the three principal players involved in the
formulation of the National Military Strategy: The President, Secretary
of Defense, and Chairman of the Joint Chiefs of Staff. Naturally there
is a considerable interaction and consultation among the parties involved
in arriving at the agreed strategy. Also, Congress is being kept informed
about the projected major military missions and the military force structure
through an annual report by the President. The process used to formulate
the National Military Strategy is determined by the Goldwater-Nichols Department of Defense Reorganization Act of 1986[12] which clarified the roles
and responsibilities of the Military Services in support of the Secretary of
Defense.

One of the provisions of the Act was to establish a larger and more
pivotal role of the Chairman of the Joint Chief of Staff, which was very
visible during Operations Desert Shield and Desert Storm. The Act established a policy of increased attention to the formulation of strategy and
contingency planning to ensure that "strategic planning and contingency
planning are linked to, and derive from, national security strategy, policies and objectives." More emphasis is now placed on joint planning and
the commanders of the unified and specified commands* must be consulted

---

*Unified commands involve two or more Services placed under a single commander
over the forces (land, naval, and aerospace), where service forces may be organized into
joint subordinate commands. The unified commands are the U.S. Atlantic Command,
U.S. European Command, U.S. Pacific Command, U.S. Southern Command, U.S. Central Command, U.S. Special Operations Command, U.S. Transportation Command, U.S.
Strategic Command, U.S. Space Command. The specified commands have a broad and
continuing mission and are normally composed of forces from one service. Currently only

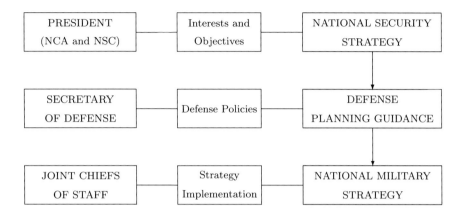

Fig. 1.2: Formulation of the National Military Strategy.

in the assessment of our warfighting capabilities. The Chairman must also advise the Secretary of Defense "on the priorities of the requirements identified by the commanders of the unified and specified combatant commands." Furthermore, the Chairman is directed to periodically (not less than biennially) review the combatant commands in terms of missions, responsibilities and force structure, and recommend changes.

The Act requires the Secretary of Defense to report annually to Congress, describing major military missions and the military force structure over the next year. The final requirement is for the President to report annually to Congress on the national security strategy and an assessment of our capabilities to support it.

The National Command Authorities (NCA) (defined as the President and the Secretary of Defense) and the National Security Council (NSC) formulate national security strategy, and integrate the domestic, foreign, and military policies related to that strategy, as shown in Fig. 1.2. This strategy must prepare for the use of military power across the entire continuum of military operations, from humanitarian assistance to shows of force, through confrontation short of war, to a range of wartime conventional or even nuclear operations.

---

one specified command is used – Forces Command, responsible for the land defense of the United States that is composed of mostly of Army forces. Another type of command is the combined command consisting of forces from more than one nation (e.g., the North Atlantic Treaty Organization (NATO) force commands).

The President signs the "National Security Strategy of the United States" report to Congress each year, as required by the Goldwater-Nichols Act. A major objective for national security stated in the 1993 document[13] is "Global and regional stability which encourages peaceful changes and progress." One subobjective is "ensuring that no hostile power is able to dominate or control a region critical to our interest."

Several documents are used to translate the national security objectives and strategies into national military objectives and strategies. The Defense Secretary publishes the "Defense Planning Guidance (DPG)," which is used by the unified and specified commands to develop required plans and provide assessment of our capabilities to meet and execute national military objectives and strategies.[14]

The Chairman of the Joint Chiefs of Staff publishes the "National Military Strategy" report,[15] which forms the basis for identification of missions, objectives, and tasks for the military, and subsequent mission area assessments and mission need statements, which are the first steps in the acquisition process for defense systems (see Chapter 2).

The current DPG[14] contains the following four essential elements of the Regional Defense Planning Strategy:

- Strategic Deterrence and Defense.
- Forward Presence.
- Crisis Response.
- Reconstitution.

These elements reflect today's changed global political and economic environment facing the U.S., and are supported six "pillars" of defense resources underlying the National Military Strategy, such as[16]

- Readiness.
- Force Structure.
- Sustainability.
- Science and Technology.
- System Acquisition.
- Infrastructure and Overhead.

These "pillars" must be considered in the planning process by the Chairman of the Joint Chiefs of Staff and the regional (unified command) commanders to establish regional objectives, and campaign operational plans and tasks.

1.4 Military Roles and Missions

Before we move to the subject of specific military objectives and tasks, it is appropriate to discuss briefly the roles and missions of the Army,

Navy, Air Force, and the Marine Corps. Roles define the broad purposes or functions of land, naval, or aerospace forces. Missions define specific tasks, not capabilities or organizations. The roles and missions are, in turn, defined by objectives, not by the weapon platform or weapon. It should be noted that most forces can perform multiple roles, and roles and missions matchups are not exclusive, e.g., some missions can support multiple roles. Table 1.1 provides a summary of the main roles and missions for the three services. These roles and missions reflect the four essential elements in the current Defense Planning Guidance (DPG): Strategic Deterrence and Defense, Forward Presence, Crisis Response, and Reconstitution.

Table 1.1: Roles and Missions of the Army, Navy, Marine Corps, and the Air Force (Missions are listed in parentheses)

U.S. Army[17]	U.S. Navy/Marines[18]	U.S. Air Force[19]
• Defeat of the Enemy Forces • Defense of Land Area • Occupation of Land Area • Battlespace Dominance (Synchronized Joint and Combined Operations) • Unconventional Operations	• Naval Expeditionary Force (Regional and Littoral Operations) • Strategic Deterrence (Strategic Attack) • Battlespace Dominance • Power Projection (Crisis Response, Forward Presence) • Force Sustainment (Sealift)	• Aerospace Control (Counterair, Counterspace) • Force Application (Strategic Attack, Interdiction, Close Air Support) • Force Enhancement (Airlift, Spacelift, Air Refueling, Electronic Combat, Surveillance and Reconnaissance, Special Operations) • Force Support (Base Operability and Defense, Logistics, Combat Support, In-Orbit Support)

1.5 Military Objectives and Tasks

Once the national military strategy has been postulated and published, the regional commands, the unified and specified combatant commands, establish regional objectives, campaign plans (including operations plans) and work with the services to develop operational tasks to support those plans. The process of developing the tasks needs to be updated at least biennially, as per Goldwater-Nichols Department of Defense Reorganization

Act, and should focus on the illustrative scenarios listed in the Defense Planning Guidance.

According to Joint Operation Planning System (JOPS) procedures, the combatant commands are responsible for planning to achieve the DPG goals and objectives.[20] The list of specific tasks to achieve successful execution of the operations plans must come from these joint planning functions. The tasks, in this regard, are the precise missions (targets) that the combatant commands postulate in the campaign plan. Each mission can then be described by the attributes and subtasks that are necessary for mission completion. The level of detail can go on and on, but the attributes must include the basic information that the soldier, sailor or airman would need to plan and complete the mission. The list would include target location and distance information, tactics, armament, supporting systems and illustrative scenario details (e.g., weather, threat and terrain profiles). The frequency of occurrence for each mission type also needs to be described. When a combatant commander lacks the resources to accomplish the task or tasks identified, a deficiency or need is identified, and the services are called upon to address that deficiency. The combatant command must first examine and evaluate alternative concepts of operation that may satisfy the identified deficiency. The services must be involved with the combatant commands in developing the task lists and options for solution in order to ensure that the CINCs* needs are being addressed. General Loh, Commander of the Air Combat Command, writes, "Obviously, the mix of choices used to develop the commander's concept of operations helped define the nature of the operational objectives and the tasks within this mission and the systems required to do the tasks."[21] Together with each combatant command, the services then clarify and standardize concept of operations, threat data, scenario environment data and the myriad of other groundrules and assumptions critical to future assessments.

There have been several good examples of accomplishing theses task lists. The joint strategic target planning system (JSTPS) is DoD-wide capability to define the specific tasks (strategic targets) under the umbrella of the strategic nuclear forces mission area. The U.S. Special Operations Command's Joint Mission Analysis (JMA) also defines specific tasks (special operations missions) for its combatant commands. Both efforts demand dedicated resources to initially compile the list and subsequently maintain it. The services are involved in these efforts to ensure that the broad joint operational needs (deficiencies) are linked to specific mission areas in order to solve any deficiency.

---

*Commanders-in-Chief

Typical resources to support this level of analysis include database management tools and campaign analysis tools. The database management tools are essential in processing the huge amount of data that will be created. Campaign analysis tools must be developed to assess each combatant command's area of responsibility in terms of exploring the operational effectiveness of alternate means of completing campaign objectives.

1.6   Mission Area Assessment

The next phase of the process is mission area assessment where the services assess their capability to complete the operational tasks under the assigned mission areas. Part of the difficulty of this phase is in the transition of needs and deficiencies from the combatant command operational categories into service mission areas. The transition is required to align the operational needs and requirements of the CINCs into the programming and budgeting processes of the services. By applying the current weapon system capabilities against assigned mission areas, the services can assess the effectiveness of the force in supporting the combatant commands. More top-level analyses are involved in this step including theater/campaign analysis, force structure and concepts of operations that often result in roadmaps and other indicators of where the services should invest acquisition resources. General Loh also wrote, "Once we understand the tasks required of us, we can determine the features, characteristics, performance, and the number of systems needed–in other words, our 'operational requirements'."[21] A Mission Needs Analysis provides the detailed information to support this approach.

1.7   Mission Needs Analysis/ Mission Need Statements

Mission needs are analyzed very specifically with respect to the effectiveness of existing systems against the specific missions on the task list. Analysis of survivability, supportability, and mission effectiveness provides the level of detail needed to support the mission area assessment. A very detailed look at the systems and subsystems may be required to identify and defend precisely where deficiencies exist and where key capabilities and characteristics exist. Mission needs analysis includes a number of supporting analyses (i.e., many-on-many, single sortie, or functional analysis, etc.), and must be frequently updated in order to provide the necessary detail to the mission area assessment, in terms of assessing the sustainability, availability, and effectiveness of our warfighting systems, and to support the justification of the mission need statement. Thus the Mission Needs Analysis and Mission

Need Statement are the major steps in the Defense Requirements Process before the Acquisition Process for defense systems begins.

References

[1] Aspin, L., and W. Dickinson, *Defense for a New Era*, Brassey's (US), Inc., Washington, DC, 1992, pp. xxv-xxvii.

[2] Nelson, Charles R., Maj., Unpublished Course Notes, Air Command and Staff College, Maxwell AFB, AL, 1991; also as "Keeping the Edge", Program Manager, Jan.-Feb. 1992, pp. 32-41.

[3] McCard, Harold K., "Research and Development Strategies," in The United States Air Force: Aerospace Challenges and Missions in the 1990s, by The International Security Studies Program, The Fletcher School of Law and Diplomacy, Tufts University, 3-4 April, 1991.

[4] Drew, Dennis M. and Donald M. Snow, "Making Strategy," Maxwell Air Force Base, AL, Air University Press, 1988, pp. 36-43.

[5] White House, "National Security Strategy of the United States," Washington, DC, Government Printing Office, 1987.

[6] *Idem*, 1988.

[7] *Idem*, 1990.

[8] *Idem*, 1991.

[9] Goldberg, Joseph E., "Strategic Success," in Essays on Strategy IV, Washington, DC, National Defense University Press, 1987.

[10] Air Force Policy Directive 10-6, "Mission Needs and Operational Requirements," Department of the Air Force, Headquarters U.S. Air Force, 19 January 1993.

[11] Air Force Instruction 10-601, "Mission Needs and Operational Requirements Guidance and Procedures," Department of the Air Force, Headquarters U.S. Air Force, 16 February 1993.

[12] Public Law 99-433, "Department of Defense Reorganization Act," United States Congress, October 1, 1986.

[13] White House, "National Security Strategy of the United States," Washington, DC, Government Printing Office, 1993.

[14] Department of Defense, The Pentagon, "Defense Planning Guidance, FY 1995-2000," Washington, DC, Government Printing Office, May 1992.

[15] Chairman, Joint Chiefs of Staff, The Pentagon, "National Military Strategy," Washington, DC, Government Printing Office, August 1992.

[16] Department of Defense, The Pentagon, "Defense Strategy for the 1990s: The Regional Defense Strategy," Washington, DC, January 1993.

[17] Department of the Army, The Pentagon, "The Army," Publication FM 100-1, Washington, DC, 10 December 1991.

[18] Department of the Navy, The Pentagon, "From the Sea: Preparing the Naval Service for the 21st Century," NNS 104, 30 September 1992.

[19] Department of the Air Force, The Pentagon, "Basic Aerospace Doctrine of the United States Air Force," AFM 1-1, Washington, DC, March 1992.

[20] Armed Forces Staff College, Pub. 1, *The Joint Staff Officer's Guide 1991*, National Defense University, Norfolk, VA, Government Printing Office, Washington, DC.

[21] Loh, Gen. John M., USAF, "Advocating Mission Needs in Tomorrow's World," Airpower Journal, Vol. 6, No. 1, Spring 1992.

Chapter 2

ACQUISITION PROCESS

2.1   History of Weapon Systems Acquisition

1940s PERIOD

System acquisition was not always as complicated as it is now. Back in 1947, when the Department of Defense (DoD) was first formed, system acquisition was a more straight-forward, simple process that could be compared to something like the automobile industry. The emphasis was on simplicity, reliability, and producibility. The DoD lacked any formal authority to control the acquisition process, having been designed to be a loose confederation of the three military departments that was designed to provide loose guidance to each department. After World War II, there was a decline in defense business, which reversed as the United States entered the Korean conflict.

1950s PERIOD

Throughout the 1950s the individual services generally ran their own acquisition programs with little interference by the Secretary of Defense, with each service buying the weapon systems it felt were suitable for the conflicts each envisioned. 1950 saw the division between research and development, and the support of weapon systems in the Air Force when the Air Materiel Command split into two commands, the Air Research and Development Command for research and development, and the Air Materiel Command to acquire and support the systems.

Defense budgets increased after the Korean conflict as a result of an increased international military role, presenting the challenges of efficiently managing the first peacetime defense industry in United States history and effectively coordinating military research and development efforts. This was an era of cost-reimbursement contracts for both development and production. The emphasis moved from an industry like the automobile industry, to an industry that was more custom design and development, where contracting played a major role. This type of emphasis continues today. The trend was towards high technology, with little emphasis on "should-cost", "design-to-cost", or "life cycle cost." Production costs did not pose a major constraint on engineering design, and when designs became impractical, they were modified in accordance with government funded engineering changes.

It was also during this period of time that the first project management organizations were born out of the need to manage the increasing complexities found in acquisition programs, and the desire to smoothly transition acquisition efforts from development to production. The project management strategy was first used on the ICBM program, and helped to push the industry to increased engineering specialization and developmental concurrency, bringing about the need for systems engineering. The Department of Defense Reorganization Act of 1958 laid the groundwork for the acquisition management structure we know today, authorizing the Secretary of Defense to assign the development, production, and operational use of weapon systems to any military department or service.

1960s PERIOD

The Department of Defense Reorganization Act of 1958 was not fully executed until 1961 when Robert McNamara became the Secretary of Defense. McNamara brought about many of the concepts that we have in our current acquisition environment. Initiatives that have persisted include: the planning, programming, and budgeting system; integrated logistics support planning; increased competition; network planning and scheduling; incentive contracting; source selection and proposal evaluation procedures; improved quality assurance; information systems; value engineering; technical data management; configuration management; the work breakdown structure; and defense standardization. McNamara also initiated the use of more paper studies, versus system prototyping, in the earlier acquisition phases as a cost savings measure.

Air Force Systems Command and the Air Force Logistics Command came into being in 1961. This change of command structure brought about the position of the program manager as we know it today, with the program manager being responsible for both the development and production of a weapon system. McNamara believed in active management from the top, and control of programs was more centralized with the DoD, which thwarted the new Systems Command's flexibility in having their program managers manage the programs they were assigned to. Contracts became more fixed price and incentive instead of cost reimbursement in response to the large cost overruns of the 1950s. Unfortunately, large cost overruns developed in a number of programs in the 1960s, programs like the C-5A, the F-111, and the SRAM-A due to numerous changes in contract requirements and the lack of enforcement of the fixed price contracts.

1970s PERIOD

The next era in acquisition history started in 1968 with a new Deputy Secretary of Defense, David Packard who believed more in decentralized control of acquisition programs, although he still felt DoD involvement was necessary in major decisions. Packard instituted the Development Concept Paper, which later became the Decision Coordinating Paper, and is now known as the Integrated Program Summary (IPS), to maintain DoD involvement. Packard also established the Defense Acquisition Review Council, which is now known as the Defense Acquisition Board, to advise him of the status of each major program prior to the program entering the next phase of development. The Cost Analysis Improvement Group was established by Packard to provide DoD with cost estimates that were independent of the program office, and to determine uniform cost estimation standards so everyone was looking at the same baseline when reviewing cost estimate information.

Packard initiated DoD Directive 5000.1, which was entitled "Acquisition of Major Defense Systems." 5000.1 was based on Packard's view that successful development, production, and deployment of major defense systems are primarily dependent on competent people, rational priorities, and clearly defined responsibilities. Packard pushed for greater authority, reward, and accountability for program managers based on this view, but received little support from the services. Another Packard initiative was the push for more hardware prototyping, which replaced McNamara's paper studies, arguing that money spent in early hardware prototyping and testing would reap cost savings in improved contractor selection. Based on the large overruns encountered under McNamara, Packard also moved the DoD back towards cost reimbursement and incentive contracting. This, unfortunately, did little to help control the cost overruns that DoD was experiencing. Packard left office in 1971, dissatisfied with the lack of success of most of his initiatives.

Other initiatives in this era included the publication of the Office of Management and Budget, or OMB, Circular A-109, entitled "Acquisition of Major Systems," in 1976. OMB Circular A-109 applied to all federal agencies, including DoD, and paralleled 5000.1 in many ways. Two areas added with OMB Circular A-109 were the emphasis on competition in the acquisition process and the need for each department to perform mission area analyses to determine their needs throughout the entire acquisition process. This circular prompted Secretary of Defense Harold Brown to add a new milestone to the acquisition process, and pushed many of the program reviews back up to the DoD level to regain some of the control that Packard had released, thus reversing the direction of control to a more centralized system. The late 1970s also saw a large defense draw-down with the Carter Administration after Vietnam.

1980s PERIOD

The 1980s started with a new administration, the Reagan Administration, that advocated a large defense build up. We had a new Secretary of Defense, Caspar Weinberger, and Deputy Secretary of Defense, Frank Carlucci who believed in a controlled, decentralized acquisition process. Carlucci developed a set of 32 initiatives designed to help streamline defense acquisition, with some of the more notable being: multi-year procurement; greater competition in contracting; stabilized budgets; more realistic budgeting; and a move to go back to fixed price contracts to help control cost overruns again. The B-1 program was restarted after having been canceled under the Carter Administration, and was supposed to have acted as a model program for many of the initiatives, and also demonstrate a better relationship with Congress.

Problems with the B-1 program, and inadequate reporting of those problems, created a backlash with Congress that caused greater micromanagement of the acquisition process. Problems like the B-1, and the rash of problems with perceived spares overpricing prompted a blue ribbon panel chaired by David Packard, which became known as the Packard Commission, in 1986 to help alleviate the acquisition problems that DoD was experiencing. The Packard Commission conducted a national survey that showed the public held the military in relatively high esteem, but found that contractors were held in relatively low esteem. The commission also found that the public believed that as much as half of the defense budget was lost to waste and fraud, that as much money was lost on fraud as it was mismanagement, and that the problems in waste and fraud in defense spending were very serious national problems of major proportion. The Packard Commission findings indicated that the defense acquisition process was not effectively managed, and they came up with four recommendations to help alleviate this problem.

The first recommendation was to create a new Under Secretary of Defense for Acquisition who would be in charge of procurement, research and development, and test and evaluation for all weapon systems. The second recommendation was to create acquisition executives in each component that reported directly to the new Under Secretary of Defense for Acquisition, as well as their Service secretary. The third recommendation was to create Program Executive Officers that would oversee specified programs within each service and report to the Component Acquisition Executive. The final recommendation was to use the Vice Chairman of the Joint Chiefs of Staff as the chairman of a new joint requirements management board to establish requirements for new systems, and to help prevent overlap of system development between the services. At the same time the Packard Commission was operating, a new bill was passed in Congress, called the Goldwater-Nichols Defense Reorganization Act of 1986, that was intended to trim headquarters staff, and to provide more inter-service coordination of defense.

Little progress, with the exception of the acquisition chain of command changes, was made on either the Packard Commission or Goldwater- Nichols initiatives until the Defense Management Review of 1989. The Defense Management Review basically agreed to implement the recommendations from both the Packard Commission and the Goldwater-Nichols bill.

1990s PERIOD

Where does this leave us now? We are seeing the effects of the Defense Management Review as you read this text. Many of the headquarters staffs have been drastically reduced, commands are being restructured, or combined, as in the case of the Air Force Systems Command and the Air Force Logistics Command combined into the new Air Force Materiel Command in July 1992. The structure that was recommended by the Packard Commission is now in place and functioning, although it can still be worked around as evidenced by the recent cancellation of the Navy A-12 program due to less than full disclosure on several agencies' parts through the streamlined acquisition reporting channels. DoD Directive (DoDD) 5000.1, "Defense Acquisition," was recently updated and released in February 1991.[1] Companion documents DoD Instruction (DoDI) 5000.2 "Defense Management Policies and Procedures," and DoD Manual (DoDM) 5000.2-M, "Defense Acquisition Management Documentation and Reports," were also published at the same time.[2,3] With this new acquisition process, we have moved back to an environment of cost type contracting due to continued cost and schedule overruns, and Congress will probably continue to micro-manage, especially in the light of the current Defense draw-down. In addition, through the enactment of the National Defense Authorization Act for Fiscal Year 1991 and the Defense Acquisition Workforce Improvement Act, Congress has mandated professionalism in the Acquisition Corps.

LESSONS FROM HISTORY

Acquisition management has been in a constant state of flux. One significant observation over the past fifty years is the cyclical nature of moving from cost to fixed price contracting, centralized control to decentralized control, and separated commands to unified commands for acquisition and support. This cyclical nature of change with the same problems persisting begs the question of whether or not we have really identified the root causes of the problems, or if we just get fed up with the problems and try and change things to see if the problems will go away. We don't know where we will be in this process in the next decade, but the one thing that can be guaranteed is that the acquisition climate will continue to change, and the ability to adapt to these changes will be crucial to success in program management.

2.2   Overview of the Acquisition Process*

THE ACQUISITION STRUCTURE

Major acquisition programs are managed from a structure that is separate from the normal operational chains of command. Management of major weapon system acquisitions requires efficient decision making and effective implementation. The three-tiered structure shown in Fig. 2.1 provides that streamlined structure, as used in the Air Force.

Program managers (PMs) are the individuals responsible for development and delivery schedules, and ensuring weapon systems perform as required. They are not the advocate for the program; that is the responsibility of the operating command. Program managers are also known as System Program Directors under the Integrated Weapon System Management concept. The PM reports directly to a Program Executive Officer (PEO).

Program Executive Officers (PEOs) are key middle managers tasked with direct accountability for the execution and information validation of a limited group of mission related major acquisition programs. The PEO positions were established as the command line between the Component Acquisition Executive (CAE), also known as the Service Acquisition Executive (SAE), and the PMs for major and selected acquisition programs. They devote full attention to this task and report only to the CAE. Air Force PEO mission areas are divided into six categories: Bombers/other systems, Information Systems, Tactical/Airlift Systems, Space Systems, Command/Communications/Control Systems, and Weapon Systems.

---

*Based on Ref. 4.

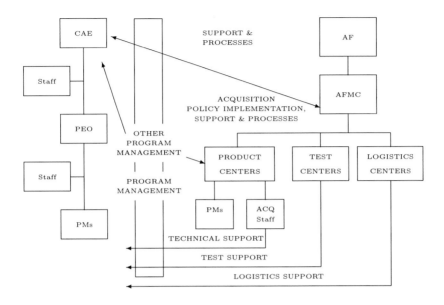

Fig. 2.1: Air Force acquisition structure.

The CAE is an assistant secretary with full-time responsibility for all service acquisition programs. The CAE, although on the staff of the Service secretary, is solely responsible to the Defense Acquisition Executive (DAE) for all Service acquisition matters. (The Secretary of the Air Force can choose to designate himself as the Acquisition Executive for a major program if he feels it necessary to do so.) The Under Secretary of Defense for Acquisition [USD(A)] has been designated the DAE. According to the Secretary of Defense's 1991 Annual Report to the President and Congress, the DAE is ultimately responsible for acquisition within the Department of Defense, having the authority for approving major defense programs at major milestones in the acquisition process, for directing the heads of the DoD components on all acquisition matters, and for directing the comptroller to withhold funds, if necessary, to ensure program milestone criteria are met.

The right side of Fig. 2.1 shows that Air Force Materiel Command (AFMC), the Air Force's command charged with weapon system acquisition, has the same organizing, training, and equipping role as the other major commands, except in this case to support service program acquisition efforts. One can also see that less than major programs are still reported to the CAE except through AFMC's Product or Logistics Centers through Designated Acquisition Commanders (DACs) who take the role of the Program Executive Officer (PEO).

Programs are assigned to an acquisition category (ACAT) based on the dollar size of, and the level of interest in, the program. ACAT I programs are Major Defense Acquisition Programs (MDAPs). An MDAP is a program that has either been designated as an MDAP by the Under Secretary of Defense for Acquisition (USD(A)), or meets dollar thresholds shown in Table 2.1. This definition is based on criteria in Title 10, United States Code, section 2430, "Major defense acquisition program defined," and reflects authorities delegated in DoDD 5134.1, "Under Secretary of Defense for Acquisition." ACAT I are further divided into ACAT ID and IC based on the level of interest in the program. Programs may move back and forth between ACAT ID and IC as they progress through the life cycle. ACAT II programs are major systems. According to the criteria in Title 10, United States Code, Section 2302, "Definitions," Subsection (5), a major system is regarded as a combination of elements that will function together to produce the capabilities required to fulfill a mission need, including hardware, equipment, software, or any combination thereof, but excluding construction or other improvements to real property. The program will normally meet the dollar thresholds shown in Table 2.1 as well to be considered ACAT II.

ACAT III and IV programs are regarded as nonmajor. According to DoDI 5000.2, "The additional distinction of acquisition categories III and IV allow DoD Component Acquisition Executives to delegate milestone decision authority to the lowest level deemed appropriate within their respective organizations." While this offers little explanation as to why there are two acquisition categories with ambiguous criteria, it does show that the general intent is to push the decision authority to the lowest possible level.

Specifying the dollar amounts prior to Milestone I is highly subjective. DoDD 5000.1 provides some "rules of thumb" for identifying a potential major program during the earliest stages of the process.[1] They are: (a) a capability that may require the use of new, leading edge technologies and an extensive development effort, (b) initiation of a major performance upgrade to an existing system that is fielded in significant quantities, or (c) when in doubt the program should be treated as if it will result in a major program (CAT I or II).

THE ACQUISITION PROCESS

Now we will shift the focus to the process of acquisition (Fig. 2.2). The acquisition process provides a logical means of progressively translating

Table 2.1: Acquisition Program Categories (ACATs)

ACAT	CRITERIA	AUTHORITY
I D	$200M RDT&E (1980) $1B Procurement	USD(A)
I C	$200M RDT&E (1980) $1B Procurement	CAE
II	$75M RDT&E (1980) $300M Procurement	CAE
III	Less than I or II	CAE Designates
IV	Less than III	CAE Designates

broadly stated mission needs into well-defined system-specific requirements. This is accomplished using an incremental commitment of resources, converting dollars into hardware. As the program progresses through the process, resource consumption increases and risk decreases.

The process of acquisition can be divided into two distinct areas: those that are considered preparatory and those that make up formal acquisition. The preparatory area of acquisition consists of the Requirements Definition Process (see Chapter 1) and the Concept Exploration and Definition (CE) phase. The formal acquisition area of the process consists of the Demonstration and Validation (DEM/VAL) phase, the Engineering and Manufacturing Development (EMD) phase, the Production and Deployment phase, and the Operations and Support phase.

Although Concept Exploration and Definition (CE) is commonly considered part of the acquisition process, a program is not formally established until Milestone I. Looking at this division between preparatory and formal acquisition helps to remind us of that. This division is not meant to imply that the preparatory activities are any less important than the formal activities. In reality, if the preparatory activities are not properly accomplished, there will be numerous problems during the formal part of the acquisition process.

The formal acquisition process really begins with the Milestone I decision at the end of CE. This is the first time that alternative concepts are identified well enough to allow development and scrutiny.

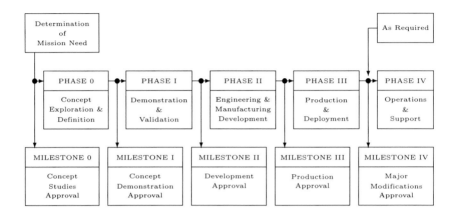

Fig. 2.2: System acquisition life cycle process.

2.3   Preparatory Acquisition Activities*

REQUIREMENTS DEFINITION

The Requirements Definition process precedes the Milestone 0 decision. It begins with an examination of the operational need which leads to trade-offs in cost, schedule and performance to determine the optimum system characteristics in later phases of the life cycle. Operational needs are generated from four process activities: mission area analysis, changes in policy, cost reduction, and taking advantage of a technological opportunity. For example, mission area analysis might reveal a threat for which we have no capability to counter. The projected improvements in Soviet radars, surface-to-air missiles, interceptors, and AWACS prompted the Air Force in the 1970s to purchase the B-1 as its survivable penetrating bomber.

Changes in policy can also result in a threat that is not countered. A good example is when President Carter canceled the B- 1. The cancellation forced the Air Force to address the increased Soviet defensive threat through support for the alternative Cruise Missile program.[5]

Cost reductions can also generate operational needs. While this deficiency usually does not stand alone as a reason for undertaking an acquisition effort, there are very few requirements identified that do not emphasize

---

*Based on Ref. 4.

cost reduction. An example is the development of the F-110 engine. While more cost efficient to operate that the F-100, it also provides a significant increase in reliability.

The opportunity offered by a technological breakthrough can lead to identification of a need. The development of stealth technology offered the possibility of decreasing pilot risk while increasing the potential for mission success. The success of the F-117 fighters during Desert Storm is ample testimony of the potential offered by taking advantage of new technology. Technology can also offer opportunities to reduce the cost of ownership or improve the effectiveness of current materiel, as demonstrated by replacing the analog flight control system in the F-111 with a digital control system to improve flight safety and increase reliability.

Any one, or all, of these deficiencies can result in a Major Command preparing a mission need statement (MNS). The MNS states the need in broad, operational terms so we are not locked into a single solution. The MNS, when completed, is sent to the Joint Requirements Oversight Council (JROC)* if it appears that the potential solution will be of sufficient magnitude to result in a new major program.

The JROC, which is chaired by the Vice Chairman of the Joints Chiefs of Staff, validates the need, considers joint service requirements that might also be met, and assigns the proposed acquisition a priority. Approximately once a year, JROC approved Mission Need Statements are considered by the Defense Acquisition Board (DAB)[†] . This board constitutes the start of the Milestone 0 - Concept Studies Approval decision process. The objectives of the Milestone 0 review are to determine if the documented mission need warrant the initiation of study efforts of alternative concepts and to define the minimum set of alternative concepts to be studied to meet the need. The DAB will determine if the need is based on a validated threat , if need cannot be satisfied by a non-materiel solution, and if the need is sufficiently important to warrant funding. If after reviewing the requirement, the DAB and the Under Secretary of Defense for Acquisition [USD(A)] concur with its validity, the USD(A) issues an Acquisition Decision Memorandum (ADM). This ADM is sent to the Service office responsible for acquisition. Funding for this phase is provided by the submitting MAJ-COM or a central studies fund controlled by the USD(A), or both.

---

*Membership of JROC consists of Vice Chairman of JCS, Vice Chiefs of Staff (Army and Air Force), Vice Chief of Naval Operations, and Commandant of the Marine Corps

[†]Membership of DAB consists of Under Secretary of Defense for Acquisition (Chair), Vice Chairman of JCS (Vice Chair), Service Acquisition Executives (Army, Navy, Air Force), Defense Comptroller, Assistant Secretary of Defense (Program Analysis and Evaluation), Director of Defense Research and Engineering, and Director of Operational Test and Evaluation

The same general principles described in the preceding paragraph apply to ACAT II-IV programs, only the level of approval authority drops commensurate with the ACAT of the program. Programs in ACAT II-IV stay within their respective components, although their Mission Need Statements (MNDs) are still supposed to be submitted to the Joint Staff to look for joint possibilities.

PHASE 0 — CONCEPT EXPLORATION AND DEFINITION

The issuance of the Milestone 0 ADM initiates Phase 0 – Concept Exploration and Definition. The emphasis during this phase is primarily on paper studies of alternatives. During this phase, the operating command initiating the Mission Need Statement (MNS) leads the study effort, establishes a concept action group to explore materiel alternatives, accomplishes a Cost and Operational Effectiveness Analysis (COEA), and prepares a brief Operational Requirements Document (ORD) with accompanying Requirements Correlation Matrix (RCM).

The COEA provides an analytical basis to support milestone decision reviews. It allows comparison of each alternative solution on the basis of cost and operational effectiveness. The ORD is a formatted statement containing performance (operational effectiveness and suitability) and related operational parameters for the proposed concept or system. An ORD is required for each concept that is proposed to advance to the next phase. The RCM is an Air Force mandatory attachment to the ORD and tracks the evolution of mission/system characteristics, parameters, or criteria that are identified in the ORD.

During the latter part of this phase, the implementing command (e.g., Air Force Materiel Command) appoints a Program Manager (PM) to establish the systems program office (SPO) cadre and begin preparing the acquisition strategy, program management plans, and the Acquisition Program Baseline (APB) for the Milestone I review.

The acquisition strategy and the program management plans provide the information essential for milestone decisions. Acquisition strategy focuses on events, explicitly linking major contractual goals and milestone decisions with development and testing. Topics addressed in the acquisition strategy include competitive prototyping, competitive alternative development and production, live fire testing as appropriate, and estimated low-rate initial production quantities, to name a few. Program plans provide for integrated, concurrent engineering of the system and its manufacturing, test, and support processes. The Acquisition Program Baseline (APB) document identifies proposed cost, schedule, and performance parameters which establish the "contract" between the Program Manager (PM) and the milestone decision authority.

In coordination with the operating command, the systems program office (SPO) develops the system maturity matrix, the Integrated Program Summary (IPS), the systems threat assessment report, and the Test and Evaluation Master Plan (TEMP) for the preferred solutions identified.

The IPS is the document that summarizes the results of the phase activities and communicates the preferred choice to the milestone decision authority. An IPS is prepared at the end of every phase and initiates the milestone review process. The TEMP helps to ensure that testing of the proposed system is well thought out and planned. It addresses test and evaluation for the entire program, including Development Test and Evaluation (DT&E) and Operational Test and Evaluation (OT&E). The TEMP begins identifying test approaches, facility requirements, and required resources.

When all of the alternatives have been evaluated the PM prepares an IPS, attaches supporting documentation prepared during Concept Exploration and Definition (CE), submits it through the PEO and CAE for JROC and DAB review. This initiates the Milestone I - Concept Demonstration Approval decision process that signals Phase 0 is complete, and the program is ready to go to Phase I. At this decision point, the DAB will review the threat assessment for validity and confirm that phase results warrant proceeding to the next phase. They will also analyze the Concept Baseline, the first acquisition program baseline, that provides initial cost, schedule, and performance objectives for Phase I; verify completion of an environmental assessment with potential mitigating plans; confirm life cycle costs and annual funding requirements are affordable, and adequate resources are programmed in the BPPBS.

2.4   Formal Acquisition Process*

PHASE I — DEMONSTRATION AND VALIDATION

If the DAE agrees with the PM's and DAB's assessment, the DAE issues a Milestone I Acquisition Decision Memorandum (ADM) authorizing start of Phase I - Demonstration and Validation (DEM/VAL). This ADM approves the acquisition strategy and Concept Baseline, establishes the requirements for the phase, and provides cost/affordability constraints.

DEM/VAL further explores the most promising solutions identified in CE. For major acquisitions, usually two competitive designs are considered

---

*Based on Ref. 4.

in order to benefit from the advantages of competition. The competition and fly-off between the Lockheed, General Dynamics, and Boeing F-22 and the Northrop and McDonald Douglas F-23 provides an excellent example of using competition to obtain high quality systems at the lowest price possible.

The objectives of Phase I are to prove critical technologies and processes are understood; better define system characteristics and capabilities; establish a proposed Development Baseline containing better defined cost, schedule, and performance estimates; and identify the preferred design approach, or best solution to the identified need. Cost generally restricts the final choice to a single system. (Eleven F-22 aircraft will be purchased during EMD for flight and stress testing. At approximately $144 million a copy it was cost prohibitive to take both aircraft into the next phase (Engineering and Manufacturing Development)). The cost for carrying two contractors and purchasing 22 aircraft for testing would have added an additional $1.5 billion.[6]

In addition to updating the documents initiated in the preceding phase, the SPO proves technical feasibility and reduces program risks; puts together a Development Baseline detailing proposed cost, schedule, and performance objectives (the second acquisition program baseline); proposes low-rate initial production quantities, addresses live-fire testing requirements; and determines the applicability of using competitive alternative development and production. Competitive alternative development and production is the development of another contractor for full scale development and production.

Prototyping and test and evaluation are used to demonstrate and validate the concept. The testing emphasis during this phase is almost completely on Development Test and Evaluation (DT&E). Operational Test and Evaluation (OT&E) efforts are usually limited to further definition of test approaches, schedules, facility requirements, and required resources.

When the above activities have been met and the PM believes the program is ready to go to the Engineering and Manufacturing Development Phase, an IPS is prepared, which, in-turn, triggers the Milestone II - Development Approval decision review. The DAB thoroughly reviews program accomplishments at this time because from this point on significant resources will be committed. They review the threat for validation, the DT&E test results to confirm that technologies and processes are attainable, the acceptability of the environmental assessment, the affordability of the life-cycle and support costs, and that programmed and budgeted resources (people and funding) are adequate.

PHASE II — ENGINEERING AND MANUFACTURING DEVELOPMENT

The DAE approves the proposed updated acquisition strategy and Development Baseline, and the Engineering and Manufacturing Development (EMD) phase begins with the issuance of the Milestone II ADM. The ADM will baseline low-rate initial production quantities, and specific cost, schedule, and performance criteria to be achieved.

The objectives of the EMD phase are to translate the design approach from DEM/VAL into a stable system design, validate the manufacturing/production processes, and demonstrate that the system produced will meet contract specifications and satisfy minimum acceptable operational performance requirements. In this phase the weapon system design is scaled-up to full size from the scaled down prototype. Full scale development often presents difficult and technically complex challenges. The F-111 is a case in point. Although the engine/inlet design performed well in subscale testing, there were substantial air flow and cavitation problems when it was constructed to full scale. The plane would not taxi without cavitating at least one of the engines in a turn. The problem was so difficult to correct that the Air Force finally had to accept the aircraft, even though it did not meet operational requirements, and compromised its employment scenarios.

During this phase the SPO will revalidate the threat, test the design under as realistic operational conditions as possible, and refine the acquisition strategy and system cost estimates. They will also develop a Production Baseline that better portrays program cost, schedule, and performance objectives (the final acquisition program baseline); develop a System Configuration Baseline; complete the environmental impact assessment; and confirm that the life-cycle and annual operational costs are affordable.

Major programs entering this phase, because of the magnitude of the resources expended, receive a tremendous amount of attention from Congress, the Office of Management and Budget, the Office of the Secretary of Defense (OSD), and the Service Chiefs. This interest makes support and advocacy by the operating command a necessity.

The operating command, because of the maturity of the system is able to begin integrating the weapon into their strategy and tactics. There will also be an increase in logistics activities as the design of the support systems are finished and acquired, and training, reliability, maintainability, and manpower requirement estimates are refined.

Like DEM/VAL, testing composes a major part of the EMD phase effort. This testing, however, is not strictly DT&E, but a combination of DT&E and Initial Operational Test and Evaluation (IOT&E). The major portion of the testing effort is on IOT&E because it must be completed and reported to the Director of Operational Test and Evaluation at OSD before a DAB can be convened. The test articles used for this testing are obtained from the low-rate initial production approved at Milestone II. The production of these articles also helps prove the manufacturing process.

The EMD phase also marks a significant change in the life of the weapons program. Once the program has made it through Milestone II and into EMD, it takes on a life of its own. As design details are completed and the system readied for production, the program will begin to consume enormous resources. These resources translate into jobs not only at the prime contractor, but also at a veritable army of subcontractors, vendors, and suppliers. The presence of these jobs in turn creates a certain level of political support in Congress. This is one reason a program that successfully enters EMD becomes extremely difficult to cancel.[7]

When the program manager has demonstrated that all technical, operational, and funding thresholds have been met, an IPS will again be prepared and submitted to trigger a DAB review. The Milestone III - Production Approval decision review will evaluate the results of Phase II for acceptable completion and ratify the costs, schedule, and performance objectives provided in the proposed Production Baseline. Once this has been completed to the satisfaction of the DAB and DAE, an ADM will be issued.

PHASE III — PRODUCTION AND DEPLOYMENT

The Production and Deployment Phase begins with the issuance of the approving ADM and its subsequent Program Management Directive (PMD). The objectives of this phase are weapon system quality and performance. In the production phase, the system is produced in quantity using assembly line methods and/fielded in large numbers. Trying to keep stable production rates in the face of annual budget perturbations becomes a major challenge.

Because production represents an even larger expense than EMD, there is great emphasis on acceptance testing of production line items and quality assurance methods to ensure production items meet article standards. Follow-on OT&E (FOT&E) will assure production articles meet effectiveness and suitability thresholds. During this phase, user personnel will train extensively to learn how to operate and maintain the system. Throughout the phase, logistics support plans are implemented and tracked to assure the system is properly supported.

Also during Phase III, improvements not incorporated in the original design will be scheduled into future production lots to minimize delaying or impacting the contractor. Problems or unplanned changes in this phase are extremely costly, often impacting other areas of production, and resulting in slips in the delivery schedule.

PHASE IV — OPERATIONS AND SUPPORT

The Operations and Support phase is really a continuation of Phase III. This phase begins after the system is fielded and has as its objectives to correct quality and safety problems, ensure the system continues to meet the threat, and identify shortcomings and deficiencies. Weapon systems are usually designed with a finite life. A deficiency results when a system will no longer sufficiently do the job due to an improved threat, a change in policy, prohibitive operational costs, technological obsolescence, or something as simple as aging. To address this deficiency one can possibly make changes in operations, maintenance, or training.

If, however, these options are found to be insufficient, and the system is still in production, the PM, along with the Operating Command, will initiate an IPS and prepare proposed options, an acquisition strategy and baseline. These actions initiate the Major Modification Approval decision process; the newest milestone decision point. These options will then be submitted to the DAB and DAE for a decision on the acquisition phase in which the program will commence and approval of the proposed acquisition strategy and baseline.

If the system is out of production a MNS will be prepared by the Operating Command that will then compete with other proposed programs in the requirements identification process. The generation of the MNS, in essence, completes the tie between Phase IV and the Requirements Identification Process for a system modification or a new system start.

The preparatory and formal acquisition process (Phases 0 through IV) is summarized in Tables 2.2, 2.3, and 2.4, which list (1) objectives, decision criteria, and acquisition decision memorandum content for each acquisition milestone, and (2) objectives and minimum required accomplishments for each acquisition phase. This can be used as a quick reference for tracking the overall process.

Table 2.2: Summary of the DoD Acquisition Process

MILESTONE 0 - Concept Studies Approval	
Objectives:	• Determine if mission need warrants further study. • Identify minimum number of study alternatives.
Decision criteria:	• Based on validated projected threat. • Non-materiel solution is unsatisfactory. • Important enough to fund for study.
Acquisition Decision Memorandum:	• Minimum number of alternatives. • Identify lead organization. • Establish exit criteria. • Identify funding.
PHASE 0 - Concept Exploration and Definition	
Objectives:	• Evaluate various alternatives. • Define most promising concept(s). • Develop supporting analyses, including risk and risk management. • Develop acquisition strategy and objectives.
Minimum required accomplishments:	• Validated threat assessment. • Identify environmental consequences. • Pros and cons of each alternative. • Proposed acquisition strategy. • Accomplish exit criteria.
MILESTONE I - Concept Demonstration Approval	
Objectives:	• Determine if Phase 0 results warrant starting a new acquisition program. • Establish the concept baseline.
Decision criteria:	• Validated threat and performance. • Environmental consequences. • Affordable life-cycle costs. • Adequate resources available.
Acquisition Decision Memorandum:	• Approve new program start. • Approve concept baseline and strategy. • Establish exit criteria. • Identify affordability constraints.

Table 2.3: Summary of the DoD Acquisition Process

PHASE I - Demonstration and Validation	
Objectives:	• Better define design and capabilities. • Demonstrate critical technologies. • Prove critical processes are attainable. • Develop supporting analyses for a Milestone II decision.
Minimum required accomplishments:	• Validated threat assessment. • Identify environmental consequences. • Identify trade-off opportunities. • Proposed development baseline. • Refine acquisition strategy. • Assess defense industrial base. • Program adequate resources. • Accomplish exit criteria.
MILESTONE II - Development Approval	
Objectives:	• Determine if Phase I results warrant continuation. • Establish the development baseline.
Decision criteria:	• Validated threat and performance. • Technologies are achievable. • Environmental consequences. • Affordable life-cycle costs. • Adequate resources available.
Acquisition Decision Memorandum:	• Approve entry into Phase II. • Approve development baseline. • Establish exit criteria. • Identify LRIP quantities.
PHASE II - Engineering and Manufacturing Development	
Objectives:	• Develop stable design. • Validate manufacturing/production processes. • Test system capabilities against mission need and specification requirements.
Minimum required accomplishments:	• Validated threat assessment. • Identify environmental consequences. • Realistic test results. • Production and configuration baselines. • Refine acquisition strategy. • Assess defense industrial base. • Program adequate resources. • Accomplish exit criteria.

Table 2.4: Summary of the DoD Acquisition Process

MILESTONE III - Production Approval	
Objectives:	• Determine if Phase II results warrant continuation.
	• Establish the production baseline.
Decision criteria:	• Validated threat and performance.
	• Reasonable test and producibility results.
	• Environmental consequences.
	• Affordable life-cycle costs.
	• Adequate resources available.
Acquisition Decision Memorandum:	• Approve entry into Phase III.
	• Approve production baseline.
	• Establish exit criteria, if appropriate.
PHASE III - Production and Deployment	
Objectives:	• Establish production and support base.
	• Achieve operational capability meeting user's need.
	• Conduct follow-on operational and production verification testing.
Minimum required accomplishments:	• Update configuration baselines.
	• Update/validate threat assessment.
	• Refine cost information.
	• Execute operational/support plans.
	• Identify operational/support problems.
MILESTONE IV - Major Modifications Approval (As required)	
Objectives:	• Determine if major upgrades to a system in production are warranted.
	• Establish the appropriate baseline.
Decision criteria:	• Validated threat and performance.
	• Field experience supports need.
	• Technologies achievable.
	• Environmental consequences.
	• Affordable life-cycle costs.
	• Adequate resources available.
Acquisition Decision Memorandum:	• Define phase to enter.
	• Approve appropriate baseline.
	• Establish exit criteria.
PHASE IV - Operations and Support	
Objectives:	• Ensure fielded system meets mission needs.
	• Identify deficiencies/shortcomings to correct or improve performance.
Minimum required accomplishments:	• Update configuration baselines.
	• Attain and maintain required performance characteristics and capabilities.
	• Conduct service life extensions.

2.5   The Biennial Planning, Programming, and Budgeting System

The BPPBS is the process through which the DoD identifies what needs to be funded and for how much. Like DoD manpower, operations, and logistics activities, acquisition efforts are funded through this system. Acquisition efforts are directly linked to the BPPBS in three specific ways; the program manager, the Defense Planning and Resources Board (DPRB), and the USD(A).

The PM is one direct link to the BPPBS. He makes a Program Objective Memorandum input through the Operating command for inclusion into the budget. Fiscal year funding is obtained through this process.

The second way the acquisition process is tied to the BPPBS is through the DPRB. The products of the BPPBS provide the basis for making informed decisions on affordability and allocation of other service funding requests when "balancing the checkbook." Programs, if found too expensive for the budget or too expensive for the benefit gained, can be unfunded in this process even if they have received Milestone approval.

The third way an acquisition program is linked to the BPPBS is through USD(A) membership on the DPRB. In this position the USD(A) could approve a program for progressing into the next phase and three months later cast a vote to unfund it, essentially voting to cancel the effort due to funding limitations.

Since the phases in the acquisition process can span several fiscal years, the progress of program implementation must be closely linked with the BPPBS. While products of the BPPBS provide the basis for making informed decisions on affordability and allocation of resources, there is no direct link between them and the Milestone decision making activities. Milestone decision reviews occur when the program is ready for consideration to move into the next phase, regardless of the BPPBS activity going on at that time.

2.6   Acquisition Process Definitions

Acquisition Program: DoD Directive 5000.1 describes an acquisition program as "a directed, funded effort that is designed to provide a new or improved materiel capability in response to a validated need." This definition also includes security assistance programs where we are working with a foreign country to upgrade their military capability.[1]

Weapon System: A weapon system is defined as the prime operating equipment and all of the ancillary functions that comprise the maintenance capability, training, technical orders, facilities, supplies, spares, manpower, and anything else needed to provide an operational capability.

Both of these definitions are extremely important to remember because one of the major problems we have in systems acquisition today. The "tunnel vision" we develop as we tend to focus on a piece of equipment and tend to lose sight of the system concept, leads us to make decisions that look good in the short term, but may have negative impacts with regards to the overall life cycle and system capability.

Guidance Documents: DoDD 5000.1 is a basic philosophy document showing the broad concepts of systems acquisition as well as the responsibilities of some of the major individuals and groups.[1] DoDI 5000.2, "Defense Acquisition Management Policies and Procedures," on the other hand, gives policies and procedures for conducting systems acquisitions, and its companion document DoD 5000.2-M describes the documentation required for the policies and procedures indicated in DoDI 5000.2.[2,3] One other document, OMB Circular A-109, which discusses acquisition philosophy and policies for the Federal Government, is still the same as when it was written in 1976. 5000.1 and 5000.2 are compliant with the philosophy and policies found in A-109. Services also write implementing regulations. For example, AFR 800-1 is the Air Force implementation of DoDD 5000.1, and AFR 800-2 is the implementation of DoDI 5000.2, which was renamed as AF SUP 1/5000.2.

Reporting Documentation: Previous sections mentioned several key documents found in the life cycle process, although the definitions of the documents were not expanded on. The first two documents we will discuss, the Mission Need Statement and the Operational Requirements Document, communicate the user's needs in the acquisition process. The Mission Need Statement (MNS) is the document that conveys a current deficiency in meeting the mission, and states the resulting need in broad operational terms. The format of the Mission Need Statement is found in DoD 5000.2-M, and Air Force processing of the Mission Need Statement can be found in AFPD 10-1 and AFI 10-601.[8,9] The Operational Requirements Document (ORD) is a further refinement of the requirements found in the Mission Need Statement, and is also written by the using command. It is the document that the program office uses in constructing the system specification, and is first written in support of Milestone I. Another document, the Requirements Correlation Matrix (RCM), traces the evolution of the requirements in the Operational Requirements Document as we progress through the life cycle. The Requirements Correlation Matrix is a key document used in writing the Test and Evaluation Master Plan (TEMP), which is discussed in Chapter 16.

Acquisition Program Baseline: This document provides the cost, schedule, and performance agreements between the Program Manager and the Defense Acquisition Executive or the Service Acquisition Executive, depending upon the category of the program. The term "Acquisition Program Baseline" can be somewhat confusing because the baseline has three different names, each indicating the update tied to a specific milestone review. The first Acquisition Program Baseline is at Milestone I, and is called the Concept Baseline. The second is at Milestone II, and is called the Development Baseline, while the third , known as the Production Baseline, occurs at Milestone III. These baselines are important to you because you may be providing information to your Program Manager (PM) to construct the baseline, and the PM will be depending on you to be aware of the baseline and to communicate anticipated problems in meeting the baseline.

Integrated Program Summary: This documents the program office activity during a phase and is presented at the Milestone review, while the Integrated Program Assessment is the distillation of the Integrated Program Summary after review by supporting staff at the DoD level. The Acquisition Decision Memorandum documents the results of a Milestone Review, and will have different types of direction at each Milestone. The final document is the Program Management Directive, which is issued by HQ Air Force after they receive an Acquisition Decision Memorandum. The Program Management Directive provides program direction, tasking, and funding levels to the various commands that will be involved in the acquisition process.

2.7 Comments

For over forty years we have been processing our National Objectives through international relations backed by a superior military capability. Acquisition was the alternative that allowed the United States to "keep the edge" over the Soviet Union. The acquisition process is a vital alternative within the National Security process. It is a methodical process, always trading-off cost, schedule, and performance in order to field the best weapon system to meet the threat. Each phase of the process has as its goal the reduction of risks and the greater assurance of performance as specified. Effective acquisition is absolutely critical for keeping our military capability at peak readiness and effectiveness.

This chapter examined how we initiate a system acquisition, the various phases and milestones contained in the life cycle process, the history of the acquisition process, and finally some guidance documents and key documentation used in this process. The life cycle will become the framework that we use to tie together much of the rest of this text.

References

[1] DoD Directive 5000.1, "Defense Acquisition," Office of the Secretary of Defense, Under Secretary of Defense for Acquisition, 23 February 1991.

[2] DoD Instruction 5000.2, "Defense Acquisition Management Policies and Procedures," Office of the Secretary of Defense, Under Secretary of Defense for Acquisition, 23 February 1991.

[3] DoD Manual 5000.2-M, "Defense Acquisition Management Documentation and Reports," Office of the Secretary of Defense, Under Secretary of Defense for Acquisition, 23 February 1991.

[4] Nelson, Charles R., Maj., Unpublished Course Notes, Air Command and Staff College, Air University, Maxwell AFB, AL, 1991; also as "Keeping the Edge," Program Manager, Jan.-Feb. 1992, pp.32-41.

[5] Werrell, Kenneth P., "The Evolution of the Cruise Missile," Air University Press, Maxwell Air Force Base, AL, 1985.

[6] *Skywriter*, Dayton, Ohio, 9 August 1991.

[7] Soileau, Randall L., "An Introduction to the System Acquisition Process," in Combat Support, ed. Ronald Carbon and Randall Soileau, Air Command and Staff College, Air University, Maxwell Air Force Base, AL, 1990.

[8] Air Force Policy Directive 10-6, "Mission Needs and Operational Requirements," Department of the Air Force, Headquarters U.S. Air Force, 19 January 1993.

[9] Air Force Instruction 10-601, "Mission Needs and Operational Requirements," Department of the Air Force, Headquarters U.S. Air Force, 16 February 1993.

Chapter 3

EXTERNAL/INTERNAL PROGRAM MANAGEMENT

3.1 External Program Management

INTRODUCTION

In Chapter 2 we described the acquisition process. In this chapter we will take a closer look at the groups and individuals that are involved in the acquisition process. After we have examined those groups and individuals, we will look at some of the other external players to the program office. Knowing the different external players is important because of the direct impact they have on the functions in the program office. Understanding the role and motivations of these players should naturally help to establish a better interface with, and satisfy the needs of the various groups and individuals.

While the people in the program office actually make the acquisition process happen, we have a number of external agents that we must deal with on a daily basis, many of whom will ultimately decide the fate of your program. This section examines the various external program interfaces and also consider the roles various commands may assume in the acquisition process, as well as the commands that normally assume these roles.

CHAIN OF COMMAND

We will start our look at external management by first examining the acquisition management chain that goes from the highest level and eventually connects back into the program office, looking at the duties of each individual as we go. In 1986, the Packard Commission found that government program management was not effective, and several recommendations were made to restructure the chain of command for acquisition programs. The Defense Management Review brought about these recommended changes,

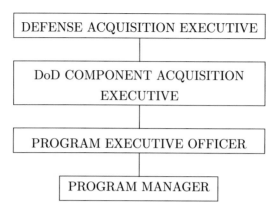

Fig. 3.1: Acquisition chain of command.

cutting out many of the layers of bureaucracy that previously existed, providing us the Chain of Command we see in Fig. 3.1.

Defense Acquisition Executive: We see that the top person in the chain of command is the Defense Acquisition Executive (DAE), whose office is the Under Secretary of Defense for Acquisition. The Defense Acquisition Executive is appointed by the Secretary of Defense, with the concurrence of Congress. His responsibility is to insure that the acquisition process runs smoothly in the Department of Defense. He accomplishes this by providing acquisition guidance to all of the Services, acting as the final authority for acquisition decisions, and by acting as the decision authority for major defense programs. In addition, the Defense Acquisition Executive is responsible for preparing long range acquisition investment area analyses and coordinating the funding of concept direction studies.

Component Acquisition Executive: The next individual in the acquisition Chain of Command is the DoD Component Acquisition Executive, also known as the Service Acquisition Executive (SAE), or, in the Air Force, as the Air Force Acquisition Executive (AFAE). The AFAE's overall responsibility is to supervise the Air Force acquisition system. AFAE does this through designating executive programs and assigning Program Executive officers, by providing direct oversight of executive programs, by chairing the Air Force Systems Acquisition Review Council, and by representing the Air Force on the Defense Acquisition Board.

Program Executive Officer: Directly below the Service Acquisition Executive is the Program Executive Officer (PEO), who is a general or Senior Executive Service person appointed by the Service Acquisition Executive. The Program Executive Officer is responsible for seeing that program direction is implemented. He accomplishes this by reviewing acquisition strategies and baselines, by approving selected management plans, financial documents, and reports, by allocating people and facilities under his control to the program offices, by interfacing with the resource allocation process on the program's behalf, and by resolving the Service Acquisition Executive programmatic issues with the program office.

The Air Force has six Program Executive Officers, as shown in Fig. 3.2. The program areas are grouped by broad categories, having the names of Tactical/Airlift; Bombers/Missiles/Trainers; Conventional Strike; Command, Control, and Communications (C3); Space; and Combat Support. The list of programs shown in Fig. 3.2 is not all inclusive, but gives a fair idea of the types of programs found in each category. All acronyms used in Fig. 3.2 are defined in Appendix A.

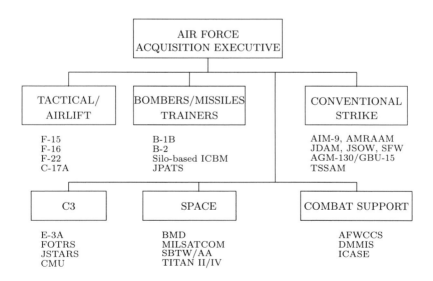

Fig. 3.2: Program Executive Officer (PEO) structure.

Program Manager: The final person, although not an external agent, in this shortened chain of command for acquisition programs is the Program Director, or Program Manager. He is selected by the component Head, and is rated by the Program Executive Officers and the Service Acquisition Executive. The program manager is the person who is responsible for the program in terms of cost, schedule, technical performance, and supportability. The program manager has numerous responsibilities that include developing an acquisition strategy, planning the program by developing a management approach, by providing budgetary estimates and alternatives, by establishing a program schedule, by developing contract strategies and structures, and conducting the day-to-day management of the program. The program manager is provided some latitude in doing this through the Acquisition Program Baseline discussed in Chapter 2.

Now that we have examined the Chain of Command, we will turn our attention to two of the external agencies that are involved in the Milestone decision process: the Joint Requirements Oversight Council, and the Defense Acquisition Board.

EXTERNAL AGENCIES

Joint Requirements Oversight Council: The Joint Requirements Oversight Council (JROC), a new council recommended by the Packard Commission in 1986, is chaired by the Vice Chairman of the Joint Chiefs of Staff. The membership is composed of all of the Vice Chiefs of Staff of all the Services. The Council has several responsibilities; validating of Mission Need Statements, prioritizing the Mission Need Statements, and reviewing programs that are coming to a Milestone review to insure that the programs are meeting their performance requirements. In validating the Mission Need Statements, the Joint Requirements Oversight Council is also looking for non-materiel solutions to the need, where there is a possibility of a joint program to meet the needs of several services, or where the need might be satisfied through a comparable allied effort. Once the need has been validated, it is assigned a priority over other competing programs from the various services, providing the opportunity for the Department of Defense to execute an overall program within a confined budget.

Defense Acquisition Board: The Defense Acquisition Board is chaired by the Under Secretary of Defense for Acquisition, also known as the Defense Acquisition Executive (DAE), and the Vice Chairman of the Joint Chiefs of Staff is the vice chair. Membership is composed of the various Component Acquisition Executives, the Defense Comptroller, the Assistant

Secretary of Defense for Program Analysis and Evaluation, the Director of Defense Research and Engineering, the Director of Operational Test and Evaluation, and the various acquisition committee chairs in the Department of Defense. This diverse membership provides the Defense Acquisition Executive a broad base of experts to judge the programs as they come before the review process.

The Defense Acquisition Board is similar to the corporate vice presidents that you would find in a major corporation. Their job is to review the Integrated Program Summaries (IPSs) and write Integrated Program Assessments (IPAs), review major programs at the milestones discussed in Chapter 2, and to recommend whether or not they should enter the next phase of the life cycle. The Defense Acquisition Board does not provide funds, or vote on the outcome of a program, rather they make recommendations to the Defense Acquisition Executive for his decision. The Defense Acquisition Board's recommendation is documented in the draft Acquisition Decision Memorandum, which is then approved or disapproved by the Defense Acquisition Executive.

Besides the external agents and agencies that are an integral part of the acquisition process, we have several other external agencies that we need to discuss in rounding out the picture of external program interfaces. These other agencies are: Congress, the General Accounting Office, the Defense Contract Audit Agency, and the Defense Contract Management Command.

Congress: Congress provides us with authorization and appropriations, both of which will be discussed in greater detail in Chapter 7, Financial Management. Congress has become more involved in the acquisition process after giving us a chance to redeem ourselves on the B-1 program, and watching us fail. We recently had another chance, and flunked again, on the A-12 program. Congress has taken to placing very specific requirements for programs in the annual authorization and appropriation legislation, causing programs to have to perform certain tasks, sometimes having to complete these low level taskings prior to having funds released. Congressional inquiries are not uncommon in program offices, and you sometimes have as little as several hours to draft a response. The Selected Acquisition Report is used to report annually to Congress on the status of major acquisition programs.

Selected Acquisition Report: The Selected Acquisition Report (SAR), provides Congress with a summary of key cost, schedule, and technical performance baselines, as well as a variance analysis from the first Selected Acquisition Report that was submitted from the program. Congress also receives several other reports on a periodic basis, and the reader is referred to DoD Instruction 5000.2, Chapter 11 for a complete listing.

General Accounting Office: Congress's investigative branch is the General Accounting Office (GAO). The GAO will end up making numerous inquiries into programs for any number of reasons. These inquiries are different from those of the Congressional staffers we just discussed. Normally the GAO will be looking into a particular issue and will actually conduct huge program audits to find the information they are looking for. People working in a program office, may find themselves providing information to the General Accounting Office. They must provide accurate answers to the investigator's questions, and then be prepared to back-brief the program office on the nature of the questions asked and responses given.

Defense Contract Audit Agency: The Defense Contract Audit Agency is concerned with auditing contracts to insure the cost reporting system is adequate, and that the bookkeeping has been properly accomplished. They will generally deal with personnel in the program control portion of the program office in performing their audits.

Defense Contract Management Command: The final agency that should be mentioned at this time is the Defense Contract Management Command. These are the people that live in the contractor's facilities and help the program office manage the day-to-day activities on a program. The Defense Contract Management Command was recently formed from all of the contract administration services, such as the AFPROs and DCASPROs. We will discuss this organization in greater detail when we talk about Post Award Contract Administration.

In addition to the external agents we have already discussed, there are several external agents within the Air Force, that are not in the Chain of Command, that we need to discuss: the Program Element Monitor and the Air Force Systems Acquisition Review Council.

Program Element Monitor: The Program Element Monitor (PEM), is an individual, in either the Secretariat or on the Air Staff, that acts as the program link with the Headquarters and Congress. The Program Element Monitor has several responsibilities that include: preparing data sheets and descriptive summaries for high level briefings, reviewing Congressional testimony, answering questions for record to Congress, and preparing the Program Management Directive we discussed earlier. It is important to know your Program Element Monitor and to keep him advised of the latest information in your program. A Program Element Monitor normally has more than one program that he manages, and by working well with the Program Element Monitor, one can help ensure a better chance of receiving priority treatment. The Program Element Monitor will normally be attending major program reviews, and this is a good place to get to know him. Again, information you manage can have a great impact on the Program Element Monitor's job.

Air Force Systems Acquisition Review Council: The Air Force Systems Acquisition Council (AFSARC) is the Air Force equivalent of the Defense Acquisition Board. The Air Force Acquisition Executive (AFAE) is the chairman, and the Board membership is comprised of all the Air Force Deputy Chiefs of Staff to ensure that all facets of implementing a new system are understood and coordinated. The Air Force Acquisition Council does the same type of things that the Defense Acquisition Board does, only for programs that are designated as less than major. This Air Force board also reviews major programs that are in the process of going to a review by the Defense Acquisition Board.

COMMAND ROLES

Now that we have discussed the external agents involved with the program in decision capacities, we will look at what commands are involved in the acquisition process as delineated in the Program Management Directive we discussed earlier The roles that commands have typically been able to assume are those of Implementing, Supporting, Operating, and Participating. In addition to these roles, different test organizations may be called out in the Program Management Directive (PMD).

Implementing Command: The first role we will discuss is that of the implementing command, which in the case of the Air Force is the recently established Air Force Materiel Command (AFMC). The implementing command supplies the personnel and facilities, in conjunction with the Program Executive Officer on major programs, to set up and run a program office. The implementing command will normally manage those programs that are in all acquisition categories and act as the Milestone Decision Authority (MDA) for programs that are in the Acquisition Category (ACAT) IV, and sometimes in ACAT III. The implementing command has responsibility for the program until the point that has been established between Phase III, Production and Deployment, and Phase IV, Operations and Support, in the acquisition life cycle, when the responsibility for the program is transferred to the supporting command.

Supporting Command: Air Force Materiel Command (AFMC) is also the supporting command. The supporting command has several responsibilities in the acquisition process, which include advising the program office of supportability considerations throughout the acquisition life cycle, planning for the necessary support equipment and facilities to maintain the weapon system when it becomes operational, and planning for when the program office moves from a Product Center, like Aeronautical Systems Center (ASC), to an Air Logistics Center, like the San Antonio Air Logistics Center.

The formation of Air Force Materiel Command (AFMC), and the establishment of Integrated Weapon System Management (IWSM), has caused a re-evaluation of the basic concepts that used to work in a System Program Office (SPO). Personnel that used to be supplied by an independent command to the program office are now in the same command as the rest of the program office. The Air Logistics Center (ALC) can be thought of as having taken the role of the supporting command, although they are also a part of Air Force Materiel Command. We will discuss more about Integrated Weapon System Management (IWSM) in Chapter 5.

Operating Command: Operating Commands are those commands that will eventually employ the systems developed by Air Force Materiel Command. Air Mobility Command (AMC), Air Combat Command (ACC), Air Force Special Operations Command (AFSOC), Air Force Intelligence Command (AFIC), Air Education and Training Command (AETC) and Air Force Space Command (AFSC) are all examples of operating commands.

The responsibilities of the operating commands are to identify operational deficiencies and opportunities, to generate the Mission Need Statements, to define the basic system requirements, and then to evolve those requirements into better defined requirements documented in the Operational Requirements Document (ORD), Operating commands work with Air Force Materiel Command to accomplish these responsibilities.

Participating Command: The final role we'll discuss is the role of the participating command. Air Education and Training Command (AETC) is sometimes designated as a participating command in systems acquisition. The participating command is responsible for carrying out the direction they have in the Program Management Directive (PMD). The participating command must also program the resources necessary to support the weapon system as it becomes operational. Although their participation may not be as in depth as some of the other commands, the participating command can have some dramatic impact on the program, especially in the training arena, and has to be afforded the same opportunity to have their needs met that the other commands have.

Test Organizations: Two other organizations that are mentioned in the Program Management Directive are the Air Force Operational Test and Evaluation Center, and the responsible test organization. The Air Force Operational Test and Evaluation Center (AFOTEC), is responsible for the operational testing of a weapon system as it prepares to go into production. The fact that they are an independent command helps to insure an impartial evaluation of the weapon system to insure that it meeds the operator's needs. The responsible test organization, on the other hand, is that test facility, like the Air Force Flight Test Center at Edwards Air Force Base, CA, that is the primary site of conducting system tests. The responsible test organization works in conjunction with the implementing and operating commands to provide the coordination necessary to conduct a test program. In Chapter 16, more information will be provided about both the Air Force Operational Test and Evaluation Center and the responsible test organization.

It is important to remember that the roles associated with the various commands are not cast in concrete. Air Force Logistics Centers, as well as all of the operating commands, perform some of their own acquisition programs. The operating command may also assume the support role of some part of the system. The key thing to remember is what the responsibilities are in each role, regardless of the command assuming the role.

3.2   Internal Program Management

INTRODUCTION

Now that we have described some of our acquisition history, the overall life cycle process, and some of the external agents that affect the program office, we will turn our attention inward to the program office so we can start hanging some information about the various functional areas onto the framework of the life cycle process.

In order to start the translation into the functional areas, we need to know how program offices are organized, that is, how the organizational structure is set up and what its implications are. By understanding the various advantages and disadvantages of various organizational structures, one is able to interact in the program office environment and understand some of the strengths and weaknesses in a particular organization.

Not only is this information applicable to the Air Force program office, it is also applicable to the various contractors. Knowing these structures can help to understand why our contractor operates the way he does, and how to find the proper personnel to take care of the questions one needs to have answered.

No introduction to organizational structures would be complete without first taking a brief look at the definition of an organization, and also the definition of a system, which is more generic than the definition we gave you for a weapon system.

Finally, we want to introduce you to the various typical functional offices found in a program office. Typical is underlined because the functional offices are just that, typical. Each program office is unique, and may group several functions within a single office that carries a different name than what we have shown. We will try and point out where some of these different groupings might occur, but you will have to become familiar with your particular program office's functional break out to be totally effective in your job.

ORGANIZATION

The first thing that we need to do is to define what an organization is. In its simplest definition, an organization is a *social system* which is *goal-directed*, has a *deliberately structured* activity system, and has an *identifiable boundary*. We will now define each of these four concepts in greater detail.

The first concept is that of social system. The basic building block of all organizations is people, and an organization must have more than one person to exist. Organizations are normally composed of many individuals, as well as groups of individuals, that interact in patterned ways to perform the essential functions of an organization.

The second concept in our definition of organizations is goal-directed. An organization exists for a specific purpose, which is normally to attain specified goals. Without goals to achieve, an organization would eventually decline and cease to exist.

The third concept is that an organization is a deliberately structured activity system. Deliberately structured means that organizational tasks are subdivided into separate departments and activities that are designed to achieve efficiencies in the work process. This structure may be loose and informal, or it may be highly structured. Activity systems means that organizations have a specific task. Individuals within the organization undertake directed activities that mesh together to accomplish the overall task, or fulfill the purpose of the organization.

The last concept in the organization definition is that of an identifiable boundary. An organization is able to distinguish which employees and organizational elements are inside, and which are outside of itself. Organizational membership is distinct. If the organization's boundary becomes unclear to people either inside or outside the organization, then the organization is in danger. Internal employees may no longer feel committed to the organization, and outside employees may not be able to identify what transactions are needed or how to conduct them. This is an important concept in program offices due to the many external agents we deal with. If these boundaries are not clearly established, it becomes difficult to know the functions you are to perform, and the detailed work that keeps an acquisition program on track is many times left unaccomplished.

SYSTEM

Now that we have examined the definition of an organization, we must look at one other concept before we talk about the three types of organizational structures; the concept of a system. Basically, a system is a set of interrelated elements that requires inputs, transforms them, and discharges outputs to the external environment. An organizational system is a bit different than a weapon system, and hopefully this wil not provide too much confusion. You will remember from our first discussion that we defined a weapon system as all of the things necessary to provide a specific capability. The definition of system is evident in the overall definition of a weapon system in that a weapon system is a set of interrelated elements (equipment, software, personnel, facilities, and data) that has inputs (people, material, money) and transforms the inputs through requirements, tactics and procedures into outputs, or the desired capability. Organizations take in material, energy, and information from the environment in which the organization exists, and then process the inputs to transform them into outputs. The term 'throughput' is used to signify these activities. To ensure efficiency and effectiveness of the 'throughput' processes, individuals within the organization develop, coordinate, and control relationships and capitalize on their inter-dependence while operating on these particular inputs. The final element of a system is the output produced by the organization. The output that results is produced from the modified inputs taken from the environment in an attempt to accomplish the organizational goals. This output may be in the form of a product or a service (see Fig. 3.3 for a schematic representation of a typical system).

Now that we have defined both organizations and systems, we need to see how those definitions fit into the work we do in systems acquisition. We are all familiar with what an organization is from our belonging to

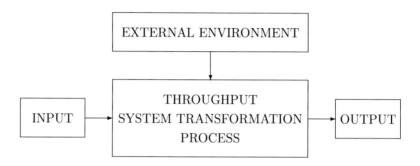

Fig. 3.3: Schematic representation of a system.

an organization, but what are the inputs, processes and outputs that our organizations deal with? Typically, our organizations have inputs of people, money, and data that is transformed, through a set of requirements and management and engineering processes, into decisions, and eventually a fielded capability. Now that we have seen how the definitions of organization and system fit into our offices, we will look at the three major ways our organizations are designed to accomplish their goals: functional organizations, project organizations, and matrix organizations.

3.3   Functional Organizations[1,2]

The first type of organizational structure is the functional structure. The functional structure groups employees together according to similar tasks and resources. Fig. 3.4 shows the typical structure of a functional organization form. As you can see, functional organizations are structured around certain specialties or disciplines.

The general manager of the functional organization has all of the specialties necessary to accomplish the goals of the organization. In our case, this is performing research and development activities, or the development and production of a weapon system. All the activities are performed within the functional groups, which are headed by functional (or department) managers. We will see that there is an engineering function, a production function, a contracting function, and so on. Within each functional area

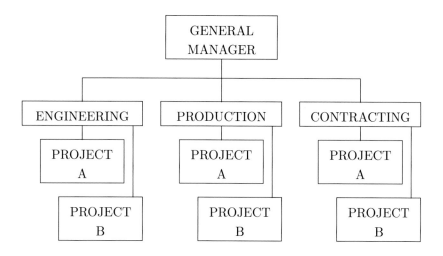

Fig. 3.4: Functional organizational structure.

are all the specialists for that particular discipline.

FUNCTIONAL ORGANIZATION ADVANTAGES

The functional organization has several distinct advantages. First, since each department maintains a strong concentration of technical expertise, and since all of the projects must 'flow' through the functional departments, each project can benefit from the most advanced technology, thus allowing for technical excellence. The functional organization, by virtue of grouping specialists together, provides an opportunity for them to share knowledge and responsibility, further enhancing the technical expertise. Second, functional organizational forms provide flexibility in the use of manpower. The functional managers are able to hire a wide variety of specialists, and any of those functional specialists are available to work on any project at the discretion of the functional manager. These personnel can even be used on many different projects if desired by the functional manager. Third, since a specialist works within a single functional department, career development and progression are usually very well defined, leaving the specialists free to concentrate on work rather than trying to figure out how they will progress up the career ladder. Fourth, functional managers maintain absolute control over their budgets. They establish their own budgets, with approval from top management, and specify the requirements for additional personnel.

Because the functional manager has flexibility in the use of the broad base of specialists, most projects can be completed within cost. Other advantages of the functional organization are that the levels of authority and responsibility are well defined, and, since each person reports to only one individual, communication channels are well structured, although they tend to be vertical, flowing up and down the chain of command. All of these advantages seem to make the functional organization appear to be a good choice.

FUNCTIONAL ORGANIZATION DISADVANTAGES

The first disadvantage of functional organizations is that functional managers tend to favor what is best for their particular interests rather than for what is best for the project. The largest or strongest functional group can tend to dominate the decision making process, and these decisions may not be consistent with, or responsive to the program objectives. A second disadvantage associated with the functional organization is related to the fact that no strong central authority or individual is responsible for the complete program being developed. As a result, coordination of activities across functional lines becomes complex, and responsibility for performance is difficult to pinpoint. The last disadvantage listed concerns the centralization of programmatic decisions to top management. Because no one program focal point exists in the functional organization, all communications needed to resolve conflicts and to make decisions must be channeled through upper management. The response to program requirements must also be filtered through several layers of management. This tends to slow down the decision making process, which can contribute to program schedule slippage. Completing all programs and tasks on time, with a high degree of quality and efficient use of resources, is all but impossible without constant involvement of top management.

FUNCTIONAL ORGANIZATION USAGE

There are four conditions that seem to favor the use of a functional organization. The first condition is that of organizational size. Functional organizations tend to work well in small to medium sized organizations that have few products, or an organizational size that is not so large and complex that coordination across functions is difficult. The second condition is the case where technology is routine, or where the primary interdependencies are within functions. The third condition that favors a functional organization is when the organization is in a stable environment. Finally, the functional organizational structure tends to work best when the organization's goals are internal efficiency, quality products, and technical specialization.

3.4   Project Organizations[1,2]

A second way of organizing our human resources is known as a project, or product, organization. This particular structure first became popular during the 1960s through its successful use in our space projects. These massive efforts, not unlike many of your weapon system programs, required coordinated efforts and activities from many functional disciplines and specialties, something that was missing in the functional organization. As shown in Fig. 3.5, the Project Organization has all of the necessary functional areas needed to complete a project grouped under a single project leader who has direct control of his resources. You will notice each project manager has a manufacturing group, an engineering group, and all the functional disciplines necessary for that particular program. All of the resources required for a program, are allocated directly to that project manager. The important aspect of the pure project organization is that the project manager owns the resources.

PROJECT ORGANIZATION ADVANTAGES

The major advantage of the pure project organization is that one individual, the project manager has responsibility for the total program and the authority over the resources necessary to accomplish the program. With this organization structure, responsiveness to program objectives is ensured. A second advantage is found in the concept of each individual reporting to only one person, thereby developing strong communication channels that result in a very rapid reaction time. A third advantage of the project organization is its ability to facilitate integration among the functional disciplines. The project manager handles all conflicts concerning the program. The project manager's role to act as the program integrator and decision maker on program trade offs is supported by having complete authority over the project.

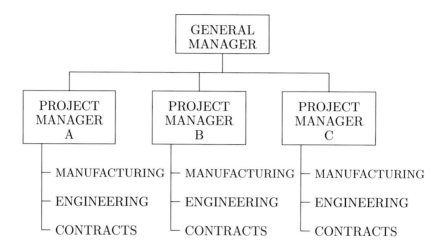

Fig. 3.5: Project organizational structure.

PROJECT ORGANIZATION DISADVANTAGES

The major disadvantage with the pure project organizational structure is the cost of maintaining the organization. There is no provision for sharing an individual with another project in order to reduce costs. For example, if you need a cost analyst to work on Project A and you only need this person about half time, you still need to hire one person. If Project B has the same situation the overall organization has two people being paid on a full time basis but only producing the work of one. The amount of slack built into this organization becomes very significant when you have multiple programs. So, if you have a particular organization office that has six programs and each project manager has their own resources, and maybe that's a half an engineer for one project, a half an engineer for another, and so on, you might have to hire six engineers when only three would be needed. A second disadvantage of the project organizational structure is the lack of well defined career progression, since there is no clear way of progressing from project to project. Another aspect of this disadvantage is a potential lack of job stability, since when one project is completed, there may or may not be a job waiting in a different project.

PROJECT ORGANIZATION USAGE

There are four conditions that lend themselves to implementing a project organizational structure. First, the organization must have sufficient resources to adequately staff a project structure. The second condition is when there are a number of interdependencies between functions in trying to implement a project. The project organizational structure facilitates communication between functional areas and puts conflict resolution at a more responsive, lower level. The third condition is when the organization exists in an unstable, and uncertain environment because of the inherent flexibility found in the project structure. Finally, a project structure is suitable when the goals of the organization are those of product specialization and innovation.

3.5   Matrix Organizations[1,2]

The final organizational structure that we will describe the matrix organizational structure, an organizational structure that implements both functional and project structures simultaneously. Fig. 3.6 shows the matrix organizational structure. You will notice that a dual hierarchy is created that affects each department in the organization. Project managers and functional managers have equal authority within the organization, and many of the employees report to both managers at the same time.

MATRIX ORGANIZATION ADVANTAGES

There are four distinct advantages to employing a matrix organizational structure. The first advantage is that it allows an organization to meet demands from more than one sector of the environment simultaneously. The matrix structure allows for the pursuit of technical excellence while maintaining strong leadership of individual projects and keeps the decision authority down to a lower level in the organization. The second advantage of the matrix structure is that it allows for the flexible use of a limited number of functional specialists. As a projects's demand for functional specialists changes, there is a ready pool of individuals to draw resources from or send resources back to. This pool of functional specialists keeps a smaller number of specialists employed that have the advantage of sharing information across projects in their functional home. The third advantage, closely related to the second advantage, is that the organization retains a great amount of flexibility in meeting changes imposed by the external environment. If a project has to start immediately, or a project is disbanded, it

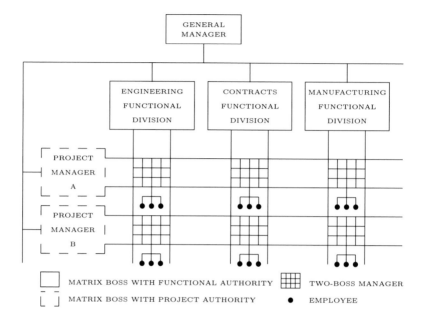

Fig. 3.6: Matrix organizational structure.

saves the attendant hiring and firing that can accompany the purely project structure, and allows more rapid response to the necessary changes. The final advantage of the matrix organization is in respect to skill development and training. There are opportunities to foster skill development in both the functional and general management areas, and forces managers to look more at the big picture in providing training and skill development.

MATRIX ORGANIZATION DISADVANTAGES

As with the other two organizational structures already examined, there are four disadvantages associated with the matrix organizational structure. The first disadvantage is the conflict and confusion that comes from having a dual authority structure. This is more commonly referred to as the "two-boss syndrome." Looking again at the matrix organizational structure, we see that the project structure is shown along the left hand side, and that the functional structure is shown along the top of the illustration. We can also see that the highlighted block shows where a two-manager boss is located.

This boss must report to both the functional manager and the project manager, which can provide a certain amount of conflict and confusion when the functional area manager and the project manager have policy disagreements that the two-boss manager has to try and resolve. This conflict and confusion can, and does flow down to the individuals located within the functional areas in a program office, helping to potentially create an overall higher level of stress in the program office. The second disadvantage of the matrix organization structure is the amount of time necessary in coordination meetings between the various projects, and between the projects and the functional home offices. The amount of time in meetings is also increased due to conflicts at the individual project employee level when the employee receives conflicting demands and direction from the project and functional leadership. The third disadvantage of the matrix structure is the additional training that must be performed. Due to the nature of the matrix, and the increased potential for conflict, courses in interpersonal relations should be taught to all personnel to better know how to handle and resolve conflict. The lack of this training, if not provided, tends to diminish the affectivity of the organization. The fourth, and final disadvantage of the matrix structure is that it is difficult to maintain a power balance between the project and functional areas. Looking again at the matrix organizational structure illustration, we can see that there is a top leader who is responsible for maintaining a balance of power between the project side and the functional side of the organization. Depending upon this leader's biases, this balance can become difficult to maintain.

MATRIX ORGANIZATION USAGE

There are again four conditions that a matrix structure is indicated for use. The first condition is where the organizational size is about medium and has several ongoing projects. Dual lines of authority become difficult to control in large organizations, and informal communication can offer sufficient coordination in small organizations. The second condition for employing the matrix structure is when there is a great amount of uncertainty in the external environment, and the environment is constantly shifting. The matrix structure provides the flexibility to surge and to cutback while still retaining valuable personnel. It also allows for both strong project and functional expertise, allowing organizations to more readily be able to capitalize on opportunities that may appear in the external environment. The third condition for employing a matrix structure is when there is a need for non-routine technology in developing a project, or when there is a high degree of interdependence between the functional areas. Finally, the fourth condition that favors the use of a matrix organization is when the organization has the dual goals of functional and project specialization. This situation is commonly found in organizations that have a number of products that are on the leading edge of technology and need the functional synergism to help them succeed.

3.6   Comparison of Functional, Project, and Matrix Organizations

Sections 3.3, 3.4, and 3.5 discussed the major features of the three types of organizations: functional, project, and matrix organizations. Table 3.1 in this section summarizes the main advantages and disadvantages of each type, and specifically points out situations when these types are appropriate to use.

3.7   Air Force Usage

Now that we have examined the three basic organizational structures, the functional, project, and matrix, how do they fit into the Air Force? Functional organizations can be found in the various laboratories that the Air Force operates, and it was historically the way that Logistics Command was organized at the Air Logistics Centers. Project structures seem to be the direction that Air Logistics Centers are now organizing. Finally, matrix organizations are generally the way Air Force Materiel Command Product Divisions are organized, as well as many of the contractors.

While it may be ideal to have a program manager decide on the type of structure that his organization should have, most often the program manager has to deal with the type of structure dictated by his parent organization. Knowing how each structure operates, its advantages and disadvantages, can help a program manager, and all of the people assigned to the program manager, know how to most effectively utilize the organizational structure they find themselves in.

PROGRAM OFFICE FUNCTIONS

The final thing we want to consider in this chapter is how a typical AF Materiel Command System Program Office (SPO) is organized. Fig. 3.7 shows the various functional areas within a typical SPO. Though the figure

Table 3.1: Comparison of Functional, Project, and Matrix Organizations

FUNCTIONAL ORGANIZATION	
ADVANTAGES:	Technical excellence Flexible personnel usage Well-defined career progression. Excellent budgetary control. Well-established communication lines.
DISADVANTAGES:	Responsiveness to project goals. No strong central project authority. Centralized decision making.
WHEN TO USE:	Small-to-medium size organizations. Routine technologies. Interdependencies in functional areas. Stable environment. Technical specialization.
PROJECT ORGANIZATION	
ADVANTAGES:	Responsiveness to project goals. Rapid reaction time. Integrated facilities between functions.
DISADVANTAGES:	Cost of maintaining organization. Ill-defined career progression.
WHEN TO USE:	Large organizational size. Interdependencies between functions. Unstable, uncertain environment. Product specialization.
MATRIX ORGANIZATION	
ADVANTAGES:	Manages dual environment demand. Flexible use of limited resources. Adaptability to environmental change. Fosters skill development and training.
DISADVANTAGES:	Conflict and confusion. Time consuming. Special training required. Difficult to maintain power balance.
WHEN TO USE:	Medium organization size. Uncertainty, shifting environment. Non-routine technology. Functional and specialization goals.

appears to be a purely project structure, most program offices are matrix organizations that gather their personnel from functional home offices. We will look at each functional area and give a basic description of what each does.

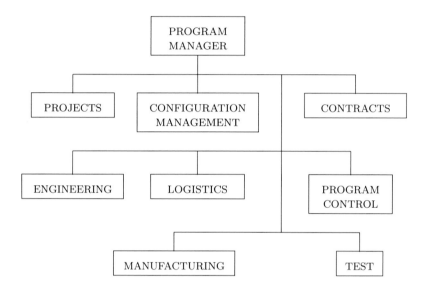

Fig. 3.7: Typical program office structure.

Projects Division: The Projects Division normally contains the various project managers that are responsible for the variety of subsystems in the total weapon system being developed. Their function is to aid the program manager in coordinating the various activities to try and keep the program within cost, schedule, performance, and support constraints. The projects division normally interacts with all of the other divisions to gather information for program manager decisions and to coordinate the activities between the other functional divisions.

Configuration Management Division: The Configuration Management Division is normally responsible for tracking the various technical baselines (the functional, allocated, and product), which we will discuss further in Chapter 11. The division also manages the engineering change proposals, and normally performs the data management functions.

Contracts Division: The Contracts Division handles all of the many contractual issues that develop in the relationship between the program and the contractor. The division is responsible for getting contract changes definitized, negotiating changes, and providing information on issues that relate to the myriad of regulations that govern the acquisition process. The division performs the only function in the program office that can officially obligate the government, and only personnel with warrants can do that.

Engineering Division: The Engineering Division is responsible for overseeing the contractor's efforts in terms of the contractor's technical efforts, verifying that the contractor is performing the technical work correctly, and insuring that the various contractor engineering specialties are working in a coordinated manner to meet the user's requirements.

Logistics Division: The Logistics Division ensures that the logistical considerations of new weapon systems are given as much priority as other considerations. This division oversees all logistical efforts and works with the gaining Air Logistics Center to make sure that their concerns are heard.

Program Control Division: The Program Control Division is normally responsible for all of the budgetary aspects of the program, maintaining a master integrated schedule, and acting as the interface with outside audit agencies. They are responsible for gathering the information to request funds, tracking the allocation of funds in the SPO, and tracking contractor cost and schedule performance.

Manufacturing Division: The Manufacturing Division is responsible for insuring that the contractor's manufacturing techniques are reasonable, that the manufacturer has the proper capability to perform the necessary work, that quality assurance is being properly implemented, and to track subcontractor performance.

Test Division: The Test Division is responsible for coordinating all of the test activities on a program, coordinating with the responsible test organization, coordinating with the independent test organization, and coordinating with the user for early operational testing.

References

[1] Meredith, Jack R. and Samuel J. Mantel, Jr., *Project Management: A Managerial Approach*, 2nd Edition, Chapter 4, pp. 112-122, John Wiley, New York, NY, 1989.

[2] Kezsbom, Deborah S., Donald L. Schilling, and Katherine A. Edward, *Dynamic Project Management: A Practical Guide for Managers and Engineers*, Chapter 2, pp. 27-37, John Wiley, New York, NY, 1989.

Chapter 4

TOTAL QUALITY (TQ)

4.1   Introduction

WHAT IS QUALITY?

Like so many other concepts, quality can be interpreted in many different ways. For those of us in the Department of Defense, certain concepts of quality are of special interest. Some of these are contained in Table 4.1.

The first statement about quality there is from the Office of Management and Budget (OMB), which speaks for the President. Notice that the OMB statement refers to quality in reference to "customer"; this is a common trait in thinking of quality and in fact most of the other statements in Table 4.1 also refer to the notion of "customer." The OMB statement also refers to "product or service"; the significance of this is that quality is broadened in scope from its traditional home in manufacturing or engineering, to now applying to literally every function, white collar as well as blue collar; in government, industry, academia, or anywhere.

The next statement in Table 4.1 is from the DoD. It introduces the notion of "correctly defined requirements." This implies that the provider or supplier of the goods or services must stay in touch with the customer; not just relying on a piece of paper that "specifies" the requirement. It suggests an "arms around" relationship between customer and supplier, not an "arms length" relationship.

The third statement is from Gen. Alfred G. Hansen, Commander of the former Air Force Logistics Command (AFLC), 1987-89. This statement is interesting in that it permits quality to "exceed" customer expectations; this should not be viewed as goldplating, rather it can be equated to "variability reduction," which results in lessened monetary losses to producers, suppliers, customers and society at large. The AFLC statement also ties quality to Reliability and Maintainability (R&M).

A statement about quality from Dr. W. Edwards Deming[1,2,3] appears next in Table 4.1. This statement includes the idea of "continually" meeting customer needs and expectations, implying that producers should antici-

61

Table 4.1: Statements About Quality

Office of Management and Budget: "The extent to which a product or service meets customer requirements and is fit for use."

Department of Defense: "Quality is conformance to correctly defined requirements satisfying customer needs."

AFLC Commander: "Quality is discipline; discipline that delivers products or services that equal or exceed customer expectations; it applies to industrial and management processes and embraces R&M. Quality is consistency in logistics process understanding, measurement and execution."

W.E. Deming:[1,2,3] "Continually meeting customers' needs and expectations at a price that they are willing to pay."

J.M. Juran:[4,5,6] "Fitness for use. Product features which meet customer needs. Freedom from product deficiencies."

Genichi Taguchi:[7,8] "The (Minimum) loss imparted by the product to society from the time the product is shipped."

P.B. Crosby:[9] "Conformance to requirements."

pate what customers might want or need; it also introduces the notion of affordability. It should be noted that price and cost are not necessarily closely related. For example, an organization may charge less than cost to attain market share, or may charge a premium for excellent quality. However, the point is, and will be made more clear later, that quality improvement drives down costs, making it more feasible to lower prices.

The next statement is from Dr. Joseph M. Juran.[4,5,6] For some years, he has been associated with the expression "fitness for use" to describe the quality of goods or services. More recently, Dr. Juran has broadened the concept of quality in the arena of "Quality Planning." In this context he introduces both a positive and a negative perspective: quality means product features which meet customer needs; and freedom from product deficiencies. "Customer" refers to anyone impacted by your process or product, internal or external to your organization.

Dr. Genichi Taguchi's view of quality is different from the others in that it is more technical.[7,8] He introduces the notion of "loss" to society, as measured in dollars. Dr. Taguchi says that quality means minimizing these losses to society.

Phil Crosby has a simple, straight-forward definition of quality: conformance to requirements. As will be seen later, this is also the old (outmoded) DoD definition.[9]

THE COST OF POOR QUALITY

Any organization has numerous ways to measure quality or the lack thereof: number of mistakes made, number of hours spent teaching, number of failures in the field, time required to complete a task, etc. How then can a top manager get motivated to implement Total Quality when there are so many indicators of quality. One way is to relate all of the indicators of poor quality to a common denominator, one that top managers are sure to understand and have an interest in. This leads to what Juran calls the language of upper management, money. This attitude-changing concept is the cost of poor quality.

The cost of poor quality is defined by Professor Resit Unal[10] as "all costs incurred to assure and assess conformance and those costs associated with non-conformance. More specifically, the cost of poor quality includes all the costs associated with poor quality including the costs aimed at preventing poor quality or assuring that quality being produced is at least acceptable."

The cost of poor quality is generally broken down into four categories: cost of prevention, cost of appraisal, cost of external failure and cost of internal failure. Each category can include any number of cost elements. For example, the cost of prevention includes cost of design review and the cost of training for quality management; cost of appraisal may include costs of implementing a Statistical Process Control (SPC) program. Internal failure costs include cost of re-work and scrap; cost of external failure includes cost of warranty repair.

Management consultants like Dr. Juran have become very successful by introducing the concept of the cost of poor quality to their clients' top management. Indeed, this concept is receiving increasing attention in the United States today. However, as organizations mature in quality management, they transcend the cost of poor quality because it ignores the most important figures. The most important figures are, in Deming's words, "invisible, unknown and unknowable." They include the cost of a dissatisfied customer, and the cost of a dissatisfied employee. Managers need to act with these invisible figures in mind. However, there are many important factors in management that can be readily measured or counted, such as payroll and bills. And these certainly can not be ignored.

The real cost of poor quality is reduced by improving quality. That is why Dr. Taguchi's monetary loss function is perhaps the best quantitative model for justifying Total Quality Management (TQM) or quality improvement. It shows that variability reduction, a form of quality improvement, drives down costs for producers, customers, and society.

QUALITY AND PRODUCTIVITY

The traditional notion that we have in our country is that quality and productivity are inversely related; in other words, as you improve quality, productivity will suffer. The modern notion of the quality-productivity relationship is that as you improve quality, productivity goes up. Was the old way of thinking wrong? Not really, but it was based on methods of quality improvement that are no longer affordable.

The traditional method of improving quality was by inspection. Do more inspecting, or inspect to tighter criteria than before and out-going quality will improve. The problem was that neither of these methods addressed the process of producing quality goods. They focused, instead, on the end product. The additional inspection increased costs. Also, it identified and sorted out more of the product as unacceptable, which increased the costs of re-work and scrap: time spent by people, time used on machines, materials, overhead, etc. All this to assure that a greater proportion of what went out the door was acceptable.

The more efficient way of improving out-going quality is by focusing on improving the processes that produce the goods or services. The processes are to be improved at least to the point where the outputs are predictable (stable) and centered on target, i.e., a very high percentage of acceptable outputs can now be predicted with confidence. The investment (not expense) in improving processes is typically not capital intensive; and as Deming points out, capital improvements should only be considered after the processes are understood and improved. The idea here is to avoid costs involved with re-write, re-work, scrap, mass inspection, etc. With this new view or paradigm of quality improvement, it can be seen that costs go down while effective capacity and productivity both increase.

BUILDING A QUALITY CULTURE

If you think back to the 17th, 18th, and early 19th centuries in America, much of what was produced was done with pride. Craftsmen, who had trained under experts, were the producers of shoes, guns, saddles, etc. They, typically, designed, built, sold, and serviced their products in small shops which they often-times owned. They lived with them. They custom-built them for the users, usually people whom they knew. Their families' reputations and pride were on the line.

Conditions changed with the advent of mechanization, interchangeable parts, and technological breakthroughs, like the transmission of electrical power. Frederick Taylor[11] popularized the concept of separating management from labor in organizations, creating large scale industrialization. Departments were created within organizations. Engineers did the planning, defined the tasks, and determined how much work should be performed. Laborers were asked only to carry out the plan, not to contribute to it. They were effectively told each day to "park your brains at the door, then come in and work."

This system was very effective in enhancing productivity from the craftsmanship era, but then, any system capitalizing on the new technologies would likely have shown improvements in at least the quantity of outputs. Initially, laborers were quite willing to participate in this system. Many were immigrants unable to speak English and wanting little more than a chance to earn and save some money, so they could someday open their own businesses in this "land of opportunity." However, they were by no means lacking common sense and intelligence.

By the nature of their jobs, these laborers were divorced from the users, had no direct contact with them. Foremen or inspectors or a quality department assumed the responsibility for the quality of the laborers' outputs. Relative to the craftsmen, laborers in the early industrial era were regarded as small, unthinking cogs in a big organization, and typically responded with much less pride of workmanship. For his contributions to this style of management, Frederick Taylor was dubbed the "Father of Scientific Management." In his later years, Taylor renounced the concept.

One person who became very concerned about the quality of goods being produced was Dr. Walter Shewhart of Bell Telephone Labs in New York in the early 1900's. He felt that the key to controlling quality, i.e., to making each unit identical to the one before it and just like the one after it, was to control process variability. To do this, Dr. Shewhart pioneered the use of statistics relative to quality and productivity.

According to Dr. Deming, Shewhart's major contributions have gone largely unnoticed. Shewhart is best known for developing a statistical tool that helps workers to assess the quality of their work, namely, the statistical quality control (SQC) chart; in fact, Shewhart is often referred to as the "Father of Statistical Quality Control." This tool can be used by a worker to determine, on a sampling basis, how well variability is being controlled, and not need to rely on an inspector. Another contribution of Dr. Shewhart was the Plan-Do-Check-Act (PDCA) cycle, which appears to be a distillation of the scientific method of problem-solving. Although many refer to the

PDCA as the "Deming Cycle," Dr. Deming modestly calls it the "Shewhart Cycle," since he learned it from Dr. Shewhart.

Deming worked under Shewhart's tutelage in the 1920's and learned the importance of variability in quality control. During World War II, a professor from Stanford University wrote to Deming, a noted statistician, asking how academia could contribute to the war effort. Deming responded that people working on military equipment should be taught about the statistical control of quality. This was accomplished. In effect, Deming took Shewhart's tool, applied it in a large scale industrial setting, and returned some of the pride of workmanship that had been lost since the era of craftsmanship in America. However, one interesting use of the concept of craftsmanship was recently begun at the San Antonio Air Logistics Center, Texas. There, workers in the Maintenance Directorate are referred to only as "craftsmen."

One of the first steps in building a *quality culture* is to understand who is the best judge of quality. While it is tempting and (perhaps inevitable) to assess our own quality, it should be realized that the customer or user is the most appropriate judge. As John Gaspari points out in his videorecording presentation *Why Quality*, we have nothing to offer our customers but quality; so they should evaluate it. Nonetheless, in the process of producing quality we can and should periodically assess our progress, even if the customer cannot participate.

BUILDING QUALITY IN SEVEN STAGES

At this time it may be useful to show a comparison of quality development in America versus. Japan. One such comparison, Table 4.2, comes from the American Supplier Institute (ASI), an organization that represents the Taguchi teachings. It suggests that the Japanese have progressed through some seven stages in the CWQC (company-wide quality control).

Table 4.2 shows that the U.S. has achieved only the first three stages, or about 40% of the Japanese level, toward TQC or Total Quality Control, as it is known in our country. Stage 1 includes the traditional quality techniques of inspection and audit; it is focused on products. The DoD has been a leader in this for many years, with the work being carried out in the field by government Quality Assurance Specialists, Quality Assurance Representatives and the like.

Stage 2 takes a more upstream view. It studies the processes and considers tools like SPC to determine process stability and capability. Recently DLA (Defense Logistics Agency) instituted an IQUE (In-plant Quality Evaluation) program which requires contractors to monitor their own

Table 4.2: The Build Up Of Quality in Seven Stages

100% JAPANESE STYLE CWQC

STAGE 7: Quality Function Deployment To Define The "Voice Of The Customer" In Operational Terms (Consumer Oriented)

STAGE 6: Quality Loss Function (Cost Oriented)

STAGE 5: Product And Process Design Optimization For More Robust Function At Lower Costs (Society Oriented)

STAGE 4: To Change The Thinking Of All Employees Through Education And Training (Humanistic)

40% U.S. STYLE TQC

STAGE 3: Quality Assurance Involving All Departments, i.e. Design, Manufacturing, Sales And Service (Systems Oriented)

STAGE 2: Quality Assurance During Production Including SPC And Foolproofing (Process Oriented)

STAGE 1: Inspection After Production, Audits Of Finished Products, And Problem Solving Activities (Product Oriented)

---

processes with less surveillance by the government; SPC is a valuable tool for these contractors.

In Stage 3 the organization takes a broader view of quality, one that envelopes all the departments in an organization, not just the traditional engineering and manufacturing groups. The DoD TQM process can be viewed as entry into Stage 3.

Stage 4 of the build up of quality involves education and training of the workforce. The managers and workers are taught the philosophy and methods involved in good management. Programs such as Total Quality (TQ) in the Air Force at the Aeronautical Systems Center, Air Force Materiel Command, exemplify this training.

Education and training can then be used in designing processes and products that will benefit mankind; concepts such as robust design, which will lead to lower costs to society, are introduced (Stages 5 and 6). The last stage introduces the concept of Quality Function Deployment (QFD), or assuring that the company is organized optimally toward satisfying the customer's perception of quality. The Air Force's R&M 2000 Variability Reduction Process (USAF/LE-RD) is examining these last three stages for applicability to the weapon systems acquisition process.

MOTIVATING FOR QUALITY

What motivates organizations to get interested in improving quality? Leadership toward excellence – sometimes. More often the answer appears to be crises – like loss of market share, survival facing American industry, or budget cuts facing the DoD. That is why Dr. Deming's most popular book entitled *Out of the Crisis* purports to describe the way out of the crisis.[2]

DEMING's CONTRIBUTIONS

Dr. W. Edwards Deming is an American whose letterhead reads "Consultant in Statistical Studies." To many people he is much more than that. He is regarded as arguably the world's foremost authority on quality. His disciples regard his philosophy of management as essential for success in today's world.[12,13]

The backbone of the Deming approach to management is a humanitarian and optimistic style of dealing with people. It also includes decision-making based on Shewhart's statistical concepts, notably the concept of a stable process, whenever possible.

It was from Dr. Shewhart of Bell Laboratories that Deming gained a great appreciation for the concept of the statistical control of quality. Shewhart recognized that the key to quality control was to reduce variability. The idea was to have every item identical to the previous one and the next one. He regarded inspection of every single item as too time consuming and expensive. It added nothing to the quality of the items.

Shewhart developed an analytical tool, based on statistical sampling concepts, that could be used by any worker to assess the amount of variability in his work process. This tool became known as the statistical quality control (SQC) chart and the use of SQC charts is now known as Statistical Process Control (SPC).

More than anything else, Dr. Deming's philosophy is captured in his *14 Obligations of Management*. These are Deming's way "out of the crisis." And they represent an answer to the question, "What do I, the top manager, need to do? What is my job?"

It should be understood that the principles included in the Deming philosophy of management were not invented or dreamed up by Dr. Deming.

They are not entirely new. But they are not in widespread use and that is the problem.

The role of a manager, according to Dr. Deming, is to create an environment and situation where his people can do a better job, not just judge. He should be a leader, facilitator, and coach. To do this the manager should be asking his subordinates "What can I do to help you to do a better job?" and mean it. This is where the practice of MBWA (management by wandering around) has its greatest value.

The concepts of the *Chain Reaction of Quality*, *Stable Processes*, the *Deadly Diseases* afflicting the Western style of management, and the *Obstacles* to quality are also vital to the Deming philosophy. They can be discussed within or outside the context of the obligations of management. In this chapter, the first two of those concepts will be discussed at the start to set the stage for the 14 points. After the 14 points, the deadly diseases and the obstacles will be covered.

The Chain Reaction of Quality:
Dr. Deming claims he would always begin his lectures to the Japanese by teaching the Chain Reaction of Quality. This begins with understanding how to improve quality. The way to improve quality is by focusing on the processes that lead to goods and services. Improve (reduce the variability) of those processes so their outputs will be predictably acceptable; this eliminates waste involved with having people, machinery, materials tied up making unacceptable product. All the costs associated with that waste are avoided.

So, the chain reaction of quality begins by understanding that improving quality drives costs down. This calls for a paradigm shift for those who believed the opposite; they thought of improving quality only by doing more inspection, or inspecting to a tighter acceptance criteria, not by reducing variability in processes.

Driving down costs by improving processes effectively increases capacity - more acceptable outcomes and fewer unacceptable outcomes will result. This means that productivity increases following quality improvements, another paradigm shift for many people.

Since quality has increased (less variability around targets) and costs have come down, the company can offer lower prices and higher quality at the same time. This leads to an increase in market share and/or an expansion of the market. The increase in market share/market assures the company of staying in business and providing more jobs. And to Dr. Deming, this is a very significant achievement.

In summary, the chain reaction says: improving quality drives down costs, which increases productivity, which results in increased market share, and that assures staying in business and providing more jobs. While no mention of "profits" is made, it is essentially assumed that this formula will result in long-term profitability. Many organizations have demonstrated this to be true.

The Concept of a Stable Process:
A stable process is one that is predictable. That is exactly what management needs - the ability to predict, with great confidence, the outputs of its processes, to aim at the right targets, and to understand its customers' needs and expectations.

The use of a statistical control chart enables one to determine whether or not a process is stable, or "in control." Greater stability means less variability, although variability may never be eliminated entirely. Managers need to understand the causes of variability. They need to know when to act on variability and what to do about it.

The 14 Obligations Of Management: The 14 points comprise a comprehensive theory of management. Deming's own discussion of them at seminars and in writing changes somewhat over time. A condensed version of the Deming's 14 points is reproduced in Table 4.3.

Table 4.3: The Deming's 14 Points Quality Message

---

1. Create constancy of purpose toward improvement of product and service, with the aim to become competitive, to stay in business and to provide jobs.

2. Adopt a new philosophy. In new economic age, Western management must awaken to the challenge, learn responsibilities and take on leadership for change.

3. Cease dependence on inspection to achieve quality. Eliminate the need for inspection on a mass basis by building quality into the product in the first place.

4. End the practice of awarding business solely on the basis of price. Instead minimize total cost. Move toward a single supplier of any one item on a long-term relationship of loyalty and trust.

5. Improve constantly and forever the system of production and service to improve quality and productivity, and thus constantly decrease costs.

6. Institute training on the job to make better use of all employees.

7. Institute leadership. The aim of leadership should be to help people and machines and gadgets to do a better job. Leadership of management is in need of overhaul, as well as leadership of production workers.

8. Drive out fear so that everyone may work effectively for the company.

9. Break down barriers between departments. People in research, design, sales and production must work as a team to foresee problems of production and in use that may be encountered with the product or service.

10. Eliminate slogan, exhortations and targets for the work force that ask for zero defects, or ask for new levels of productivity without providing methods.

11. Eliminate work standards (quotas) on the factory floor. Substitute leadership. Eliminate management by objective. Eliminate management by numbers and numerical goals. Substitute leadership.

12. Remove barriers that rob the hourly worker of his right to pride of workmanship. The responsibility of supervisors must be changed from sheer numbers to quality.

13. Institute a vigorous program of education and self-improvement.

14. Clearly define to management's permanent commitment to quality and productivity and its obligation to implement these principles. Put everybody in the company to work to accomplish the transformation. The transformation is everybody's job.

The Directorate of Distribution (DS) at the Sacramento Air Logistics Center, is using the 14 points to assist in their implementation of the Civil Service Demonstration Project and Essential Process Management (Codeword PACER SHARE). Their modified version of the fourteen points is presented in Table 4.4.

Table 4.4: PACER SHARE Version of the 14 Points

---

1. All members of DS, at all levels, commit themselves to increasing productivity through higher quality of both goods and services, improved relations with all members of DS and their representative organizations, and reduced costs of operation in order to improve our ability to provide customer service in both war time and peace time environments, and thus increase our ability to serve our ultimate customer, the American people.

2. We fully subscribe to adopt the new philosophy for economic stability and improved performance and will refuse to allow commonly accepted levels of delays, mistake, defective materials and defective workmanship.

3. We will cease dependence on mass inspection as a way to achieve quality. We will require, instead, statistical evidence of built in quality.

4. To the extent that is within our charter we will work to end the process of awarding contracts and business on the basis of price tag. Instead we will work to assure that all such awards depend on meaningful measures of quality along with price. We will work to eliminate suppliers of goods and services that cannot quality with statistical evidence of quality. We will work to assure that purchasing as a separate action is eliminated and that it is combined/integrated with design of product/service, manufacturing/operations management and planning, to work with the chosen supplier, the aim being to minimize total cost, not merely initial cost to the government.

5. All members of DS will dedicate themselves to look for problems. We shall improve constantly and forever every process and system in Distribution, and work to increase quality and productivity, reduce variation, and thus reduce costs.

6. Distribution shall institute training and education on and about the job for all members.

7. We shall develop supervision so that supervisors are capable of helping people, improving methods, materials, machines, systems and processes so as to help all members, and the organization, to do a better job.

8. Distribution management shall create an atmosphere where all members are encouraged to talk openly about their jobs, make suggestions, question procedures, participate in Quality Circles and Task Forces, etc. without being concerned that their careers or jobs are endangered or that they will lose favor with supervision and management.

9. Management will dedicate itself to removing barriers between divisions, branches, and all DS organizations. All members will work to understand that we must work as a single team to foresee, and prevent, problems that may be encountered in the conduct of our vital defense mission.

10. Distribution will eliminate slogans and targets asking for increased productivity without providing methods.

11. Distribution will eliminate all work standards that prescribe ONLY numerical quotas, without the essential quality and process/system guidance required to meet the standard.

12. The Directorate dedicates itself to removing any and all barriers that inhibit ALL members' rights, at any level, to pride of workmanship.

13. Distribution will institute and continue to support a vigorous program of education and retraining. New skills are and will be required for changes in techniques, materials and services and the directorate acknowledgers its obligation to provide training/formation to its members to gain proficiency in as many of these skills/capabilities as possible.

14. Distribution will work through its existing structure, everywhere and always to effect the change to these principles of management and, when the existing structure proves inadequate to the task, will change the structure so that support for never ending improvement can be sustained.

The Deadly Diseases:
One thing America should not try to export to a friendly country, according to Dr. Deming, is its management style. He has identified certain serious ailments afflicting the Western style of management. These are: 1) lack of constancy of purpose (the U.S. Constitution can serve as a statement of purpose for the DoD); 2) emphasis on short-term profits (creative accounting, such as delaying military pay one day into the following fiscal year is

an example of this disease); 3) evaluation of performance, merit rating, or annual review; 4) mobility of management; 5) running a company on visible figures alone. Additionally there are two that are unique to America: excessive medical costs and excessive costs of liability, particularly when contingency fees are utilized.

Besides the deadly diseases, Dr. Deming has identified obstacles, easier to cure. Some of these are more readily identifiable in an organization that is starting a quality initiative than are the deadly diseases.

Some of the obstacles (and a brief rebuttal), are: the hope for instant pudding (cultural change takes time); the supposition that solving problems or new machinery will transform industry (transformation of management style is needed); trying to copy another organization (understand why it worked there); the following notions: "it won't work here - our problems are different (your problems are different, but the obligations of management are universal)," "the quality department will take care of it" (management and the workers can contribute more to quality than the quality department), "if only the workers would work harder, our problems would disappear" (most problems are the fault of the system, which management controls), "meeting the specification is good enough (need to know more than is in the specification, e.g., how the product will be used, in order to satisfy the customer);" and false starts - like quality circles and process action teams (trying to live the 14 points is the right start).

JURAN's CONTRIBUTIONS

Joseph M. Juran is another of the world's foremost authorities on quality. He is an American management consultant, who, like Deming, contributed significantly to the Japanese in the early 1950's. Juran has been highly honored by Japan and many other countries. Dr. Juran is editor of the Quality Control Handbook, now in its fourth edition. This book is regarded by many as the "bible" of quality control.[14]

Juran has written numerous other books. He is the co-author of "Quality Planning and Analysis," which has been used in AFIT's graduate courses. Dr. Juran is also chairman emeritus of the Juran Institute, which teaches, researches, and consults for organizations around the world. Juran Institute sponsors the annual IMPRO, a conference that promotes the Juran concept of quality management. There are two contributions from Dr. Juran of particular significance for management today. These are the concepts of the quality management trilogy and quality leadership.

JURAN's QUALITY MANAGEMENT TRILOGY AND QUALITY LEADER-SHIP

According to Juran, the concept of quality management, i.e., managing for quality, is analoguous to financial management. This is important to understand because it is like a key into the thinking of upper management, which already does financial management.

Financial management consists of three basic processes: 1) planning, i.e., setting goals, objectives, and how to meet them, culminating in a budget; 2) control, also known as cost control or expense control; that is, executing the plan; and 3) profit improvement (or cost reduction). If the focus were shifted from finance to quality, the same three basic processes result in quality planning, quality control and quality improvement. These processes comprise Juran's quality management trilogy.

To illustrate, consider the next fiscal year. Quality planning would be complete prior to 1 October. It would set out various quality objectives such as improve the reliability of system XYZ by 5% next year and the means to achieve it. Quality control is simply the work done by all the workers, with the possible exception of supervisory/management work. Quality control really means executing the plan. If the plan is perfect, which it never is, and it were carried out perfectly, there would be no deficiencies or mistakes. The third process in the trilogy is quality improvement. During the year, various teams are busy solving problems, or improving critical processes.

Quality Planning:
Let's look more closely at the processes in the trilogy. First, Quality Planning. It includes the generation of quality control objectives and/or quality improvement objectives. The significance here is that work required to achieve the objectives should be budgeted for, not "taken out of hide." Moreover, the objectives should be aligned with the organization's quality policy, which in turn should be driven by its mission statement.

Quality planning is how you build quality in from the start. It involves discovering and meeting customer needs. There is a sequence of steps in Juran's concept of quality planning: identify the customer; identify customer needs; translate those needs in language of our company; determine features to meet the needs; determine goals (may be numerical) for those features; develop and optimize processes to achieve the goals; prove the processes are indeed capable under operating conditions; then, turn the plan over to the operating forces, Juran's term for those who execute the plan, i.e., who do the quality control. Juran Institute offers a two-day introductory course in quality planning, plus a longer course in implementation. Juran's text, "Quality Planning" is also helpful.

Some organizations set aside a particular month, November seems to be popular in Japan, for quality planning. They review the previous year's accomplishments against the plan and decide on the new plan. In Juran's view, quality planning should be lead by upper management, just as it leads financial planning.

Quality Control:
Quality Control, the second process in the quality management trilogy, is the oldest and most widely practiced process in the trilogy. It may be helpful to distinguish quality control from quality assurance. The American Society for Quality Control (ASQC) defines quality control as "making quality what it should be"; quality assurance is "making sure quality is what it should be." These definitions do not conflict with Juran's concept. Quality assurance may also be thought of as the gathering and analysis of information on the operations so corrective action and process improvement can be taken. One traditional notion in the DoD that is now being revisited is that the contractor is specifically charged with quality control, while the Government is responsible for quality assurance.

The aim of quality control is to carry out the plan, i.e., maintain all the quality that has been "planned in" and also to fight the fires that sporadically occur. For example, the time history of a quality characteristic may be charted; when the quality characteristic unexpectedly departs from statistical control some investigation would certainly be warranted. When the fire is successfully brought under control, things are back to "normal." This does not represent quality improvement, merely it is a return to the existing standard or level of planned (this word is used tongue-in-cheek by Juran) waste. This existing level of waste may be chronic, i.e., it may have been in existence for years. However, no one really knew how bad it was or thought much of it, because the organization was "fat, dumb and happy." In fact, many times a "fire" is recognized without the use of any hard data. It is simply felt by the organization that things are much worse than usual and something needs to be done.

Quality Improvement:
Quality improvement is the third process in the trilogy. It represents a breakthrough to a new and better level of quality. According to Juran, quality improvement occurs in one way and in one way only: project by project. A project is a "problem scheduled to be resolved." Quality improvement projects are worked by teams while others are busy executing the plan. So, quality improvement goes on at the same time as does quality control.

According to Juran there is a universal sequence of steps that lead to quality improvement. These are spelled out in his book "Managerial Breakthrough." As Juran sees it, quality improvement is a series of breakthroughs. He defines breakthrough as "the organized creation of beneficial change."

There is a universal sequence of steps in Juran's concept of quality improvement. These are: breakthrough in attitude; Pareto analysis; organization of steering and diagnostic arms; breakthrough in knowledge - diagnosis; breakthrough in cultural patterns; breakthrough in results, holding the gain.

"Breakthrough in attitude" boils down to the realization by top management that poor quality is costing them big bucks. In fact, Juran exploited this concept many years ago as a consultant. He would examine a client's operations for an extended period of time, identify the poor quality therein and try to "dollarize" its impact. Then he would reveal his findings to the top manager, who would be taken aback by the magnitude of the "gold in the mine", another expression of Juran's. The result was that the top manager's attitude about quality had changed and Juran had his foot in the door.

Once top management has bought into the notion that there's money to be saved by improving quality, the next step is to determine what quality improvements to make first. That's where the Pareto analysis comes in. As used here Pareto analysis is simply a listing, typically in descending order of importance, of all the quality problems within the organization. For example, if management considers the cost of poor quality as important, it would rank its quality problems in order of decreasing dollars. It would attack the big dollar problems first. Juran calls these the "vital few." It would put the remaining problems, the "trivial many" or "useful many" on the back burner.

What's most important in Pareto analysis is where you start. In other words management must decide what's really important to the organization. For example, it may be that instead of the cost of poor quality, what's important is: mission availability rate, user satisfaction, number of defectives delivered, timeliness of deliveries, or the number of course offerings presented. Only when management faces up to this issue, will the quality improvement team be led effectively toward attacking the "vital few."

4.2   TQ in the Federal Government

The recent upsurge in the Federal government's interest in quality management followed President Reagan's issuance of Executive Order 12552. It authorized the Office of Management and Budget (OMB) to establish productivity improvement efforts in several executive departments and agencies, including the DoD.

The objectives of OMB Circular A-132, Federal Productivity and Quality Improvement in Service Delivery, April, 1988 are to: a) implement quality and productivity management practices in every Federal agency; and b) make continuous, incremental improvements in quality, timeliness and efficiency of services.

A Federal Quality Institute (FQI) was recently established to introduce top-level federal agency personnel to the Total Quality Management (TQM) concept. Some twenty-four contractors are now authorized via a Federal Master Contract to teach TQM. They teach a variety of concepts typically including the philosophies of Deming, Juran, and Crosby.

FQI describes TQM "as being a systematic way to improve products and services, not a new program; it uses a structured approach to identifying and solving problems, not "fighting fires"; long term, not short term; conveyed by management action, not by slogans; supported by statistical quality control, not driven by it; practiced by everyone, not delegated to subordinates."

The Office of Personnel Management (OPM) also offers a series of "Quality Improvement" courses. Such courses are aimed at meeting the training needs of Executive Order 12637 (Productivity Improvement Program for the Federal Government) and OMB Circular A-132.

Other federal agencies have also "done their own thing" in training for TQM. For example, the IRS has invested heavily in training from the Juran Institute. The DoD has purchased the majority of slots at some of Dr. Deming's Seminars, and units within DOD have been getting training from a plethora of other sources, including the Air Force Institute of Technology.

HISTORY OF QUALITY EFFORTS IN THE DoD

The DoD issued a TQM Master Plan in August, 1988. The June/July 1989 DoD TQM Message stated: "TQM is both a philosophy and a set of guiding principles that represent the foundation of a continuously improving organization. TQM is the application of quantitative methods and human resources to improve the material and services supplied to an organization, all the processes within an organization and the degree to which the needs of the customer are met, now and in the future."

Various elements of DoD had already been making their own attempts to improve quality since the early 1980's, notably Navy Aviation Depots (and their predecessor Naval Aviation Repair Facilities). The Air Force

Logistics Command was also beginning its quality effort before formal DoD efforts were announced.

The DoD has not officially endorsed the philosophy or principles of any one particular quality management guru as the "way to go." At the Office of the Secretary of Defense (OSD) level there exists a very strong Deming flavor. The DoD has produced a Total Quality Management Guide (DoD 5000.1-51G, final draft, 2/15/90) consisting of two volumes: Volume I - Key Features of the DoD Implementation and Volume II - A Guide to Implementation. Volume II exhibits a very strong influence from the philosophy of Dr. Deming. On the other hand, Dr. Juran and Phil Crosby have also been invited to give talks to upper managers at the Pentagon.

Besides Dr. Deming, the other strong influence on OSD quality thinking appears to be from the U.S. Navy experience. The Navy Personnel Research and Development Center (NPRDC) prepared the "Total Quality Management Implementation: Selected Readings" published by Direction of the Office of the Assistant Secretary of Defense (P&L) TQM/IPQ, April, 1989. In 1990, NPRDC's Dr. Laurie Broedling was named Under Secretary of Defense for TQM.

NPRDC published "A Total Quality Management Process Improvement Model" in December, 1988. It is based on the Plan- Do-Check-Act (PDCA) cycle advocated by Dr. Deming and his late colleague, Dr. Shewart, recognized as the "Father of Statistical Quality Control."

In May, 1990, NPRDC published "Readings on Managing Organizational Quality." It states that the contents are consistent with the philosophy of W. Edwards Deming. It also points out that "...the DON (Department of the Navy) recently adopted the Deming approach to improving quality."

TQ EDUCATION IN THE DoD

The Defense Systems Management College has for years led the way in teaching systems acquisition program management. The DoD initially identified the acquisition community for implementation of TQM. Recently, the Defense Acquisition University (DAU) was established as a consortium of DoD and service schools to provide education, training, and research in the area of acquisition (See Chapter 20).

Air Force Institute of Technology (AFIT), a consortium member of DAU, has been teaching, consulting, and researching quality for years. Quality as a concept has been integrated into both management and engineering curricula to ensure that all AFIT graduates will be able to incorporate quality into the engineering design and acquisition of defense systems. The Air Force Quality Center (AFQC) was established in 1992 at Air University, Maxwell AFB, AL; its purpose is to establish a center of expertise in Total Quality and to oversee training and consulting for quality within the Air Force.

4.3 TQ Concept in the Air Force

The Air Force has been vigorously pursuing the total quality concepts in support of its stated *overall* mission of defending the United States through control and exploitation of air and space. All Air Force major commands participate in this activity and are committed to developing and using quality measures or metrics to assess the ongoing progress. Air Force leaders from the Quality Council on down are committed to quality and to building the world's most respected air and space force to fullfil their corporate strategic plan of global reach and global power for America. The quest for quality in the Air Force is more than a concept; it is the Air Force's credo demanding a "leadership commitment and operating style that inspires trust, teamwork, and continuous improvement everywhere in the Air Force." The quality adage, adapted by the Air Force, says it all. "If it is broken, fix it; if it is not broken, make it better." This section summarizes some of the TQ concepts that have been incorporated into the Air Force activities.

COMPONENTS OF QUALITY

The four basic components of quality used by the Air Force are:
- Complete customer focus.
- Continuous process and product improvements.
- Empowerment of employees.
- Leadership.

Complete Customer Focus:
Customer focus is perhaps the key component for success. The product or service provided to the Air Force must meet customer requirements, and it is the customer who defines the quality measured against the Air Force performance. This in turn mandates a feedback from customers and subsequent actions to improve products and services. This customer-oriented culture of "doing business" is illustrated in Table **??**, which compares the old style versus the new style of Total Quality.

Continuous Process and Product Improvement:
Products and services are improved through process reviews, which use data collected from the quality metrics to ensure that appropriate decisions are

Table 4.5: Quality Culture: Customer Focus

OLD STYLE	NEW STYLE
• Unidentified customer	• Face-to-face real person
• The Boss	• Any recipient of products
• Remote from work	• Integral part of work
• Take this product and love it	• React to the voice of the customer – listen, listen
• Assumed satisfaction	• Continuous measurable feedback
• We think we have the best products	• Customers boast about our products
• We are the best – no room for improvement	• Continuous improvement of product

made for improvements. These improvements are based on measurements and analyses that reveal trends, causes, and effects, and they point toward correct remedies. The measurements require a feedback, as illustrated in Fig. 4.1, which shows schematically the input-output relationship for a typical process involving the customer and supplier. The requirements and feedback loop is a vital link in this process.

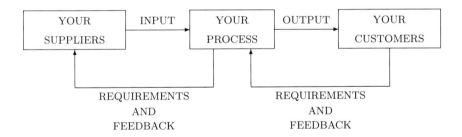

Fig. 4.1: Feedback loops for continuous process and product improvement.

Empowerment of Employees:
The key to success on this component is to create a work environment that supports employee involvement, contribution, and teamwork. Once this environment is created, the management and employees trust and respect each other, and jointly work toward continuous improvement of the prod-

uct and services they are producing. The employees have the authority to act, a process that contributes to satisfaction and self-actualization of individuals involved. It should, however, be pointed out that the improvement process is only possible if the employees are trained and educated to do their work correctly. Thus, an aggressive education and training program must complement the quality program. This is, in fact, the major thrust of the quality implementation in the Air Force today.

Leadership:
Finally, the Air Force leadership plays an important role in the Total Quality concept. This requires a leadership that is firmly committed to the program. Also, all middle-level managers must be committed to quality, must be given resources to improve quality, must remove barriers to excellence, and must emphasize importance of people. Top leadership, on the other hand, must have a strategic plan that emphasizes quality goals, which in turn are translated into quality objectives. This is being done now in the Air Force by developing a clear roadmap, consisting of goals, objectives, and tasks, for use by all Air Force people.

4.4   Congressional Study of Quality

There are not many indications that the U.S. Congress is very positive about the concept of managing for quality. Certain Congressmen criticized the education and training budget for TQ. They wanted a quick payoff and were not interested in procuring a philosophy.

There have been some positive indications. For example, the House Republican Research Committee Task Force on High Technology and Competitiveness published a 1988 study entitled "Quality as a Means to Improving our Nation's Competitiveness." It should be noted that the Task Force's steering committee included Dick Cheney, who later became Secretary of Defense.

The study pointed out that quality management and the cost of poor quality are key factors in future competitiveness of American businesses. It also said that more popular urgings, like streamlining our regulations and ensuring that foreign markets are accessible, are laudable but not sufficient. The study claimed that the annual cost of poor quality in our country is about $920 billion, whereas the production of goods and services totals $3.7 trillion. This computes to roughly 25% level of waste, similar to what the quality gurus have been saying for years.

Perhaps the most promising aspect of the Task Force's study was the conclusion that their own first step must be to "alert members of Congress, key constituents, and the nation, to the dynamic potential of a quality focus in industry and government." Additionally they pointed out that the federal government is the biggest customer and the biggest supplier in America and is thus poised to influence the private sector through changes in the contracting and procurement process.

There are also other positive signs of a quality management focus in Congress. The first annual national symposium on TQM in Academia was held in July, 1990. Congressman Don Ritter was one of the principal speakers and advocates.

4.5 DoD's Ultimate Customer

Who is the "ultimate customer" for the DoD? According to DoD publications, the ultimate customer for the acquisition community is the user in the field, the airman, soldier, sailor and marine. This should include those people who operate, fix, and maintain the weapons being developed and procured. Some would argue that the ultimate customer is the taxpayer. However, perhaps we can view the taxpayer instead as the beneficiary and financier of it all.

Besides the ultimate customer, we can consider immediate and intermediate customers. For example, an AFIT professor in the classroom can view the students as immediate customers, the recipients of knowledge and understanding; the students are also "processors" when they work with the knowledge being provided; they also become suppliers when they provide feedback in the form of comments, answers, or questions. This illustrates Dr. Juran's triple role concept, i.e., we take turns being customer, processor, supplier.

If the students in the classroom are the professor's immediate customers, those who are impacted back on the job by these students (their peers, colleagues, supervisors, etc) can be considered the professor's intermediate customers. The intent is that whatever learning occurs by the student will ultimate improve the quality of weapons and support being delivered to the user in the field.

WHAT IS THE CULTURAL CHANGE?

What cultural change does DoD envision for us? Table 4.6 is a listing of old and new thinking. It comes from the Office of Secretary of Defense Directorate on Industrial Productivity and Quality, and it provides the DoD views on quality.

Table 4.6: Two Views of Quality

Traditional View	DoD Current Posture
Productivity and quality are conflicting goals.	Productivity gains are achieved through quality improvements.
Quality defined as conformance to specifications or standards.	Quality is conformance to correctly defined requirements satisfying user needs.
Quality measured by degree of nonconformance.	Quality is measured by continuous process/product improvement and user satisfaction.
Quality is achieved through intensive product inspection.	Quality is determined by product design and is achieved by effective process controls.
Some defects are allowed if product meets minimum quality standards.	Defects are prevented through process control techniques.
Quality is a separate function and focused on evaluating production.	Quality is a part of every function in all phases of the product life cycle.
Workers are blamed for poor quality.	Management is responsible for quality.
Supplier relationships are short termed and cost oriented.	Supplier relationships are long term and quality oriented.

TQ "SUCCESS STORIES"

It is interesting to note that despite a Dr. Deming influence in the writings coming out of OSD, the implementation within DoD is clearly of a Juran flavor. That is, a top-management group is formed (Air Force uses the same expression as Juran, i.e., Quality Council, while the Navy prefers the term Executive Steering Committee); projects are identified for improvement/resolution. Teams are formed (Air Force uses the name Process Action Team; Navy counterpart is a Quality Management Board). Training in how to solve problems or improve processes is heavily invested in.

OMB has been identifying "Quality Improvement Prototypes" within the federal government since 1988. The DoD has had several such winners including: Navy Aviation Depot, Cherry Point, NC; Norfolk Navy Shipyard, Norfolk, VA; Navy Forms and Publication Center, Philadelphia, PA; and the Defense Industrial Supply Center (DISC), Philadelphia, PA; Sacramento Air Logistics Center, CA; Warner-Robins Air Logistics Center, GA; and Aeronautical Systems Center, Wright-Patterson Air Force Base, OH.

Tom Peters, celebrated writer, speaker, and TV personality completed a program on "Excellence in the Public Sector," televised in 1989.[15] It showcased five organizations including the Navy's Aviation Repair facility in Alameda, CA. The former Air Force Logistics Command won the President's Award in 1991. One notable attempt in the DoD to implement Dr. Deming's philosophy is the PACER SHARE program mentioned previously.

Additionally there are numerous experiments throughout DoD to implement TQ-related concepts. The Air Force's PALACE COMPETE Program gives top management at various locations greater liberty on how to best utilize its civilian pay budget, without some of the constraints from Washington.

4.6   Software TQ

INTRODUCTION

While total quality (TQ) has been emphasized for hardware in recent years, quality has not always been stressed for computer software. Most software has been delivered to DoD with numerous errors, many of which have been serious. Since modern weapon systems often incorporate software containing millions of lines of code, and since modern systems are totally dependent on properly-functioning software, software TQM is a requisite for most modern systems.

CURRENT DoD GUIDANCE

Unfortunately, there is a scarcity of regulations and guidance addressing software TQM. One document that can help, however, is DoD-STD-2168, "Defense System Software Quality Program". This Standard is required for inclusion in contracts for which mission critical computer software is developed for a DoD agency, and can be included in contracts for which any type of software is developed for a DoD agency.

While DoD-STD-2168 does not tell a contractor how to perform software TQM, the Standard does require certain TQM activities from a contractor. The contractor is required to establish a program for assuring the quality of deliverable software, non-deliverable software, and the processes used to produce software. The contractor's quality activities must begin early in a project and be performed by personnel independent from the software development team.

SOFTWARE QUALITY ESTIMATION AND MEASUREMENT

Perhaps the primary reason for a lack of guidance in software TQM is that software quality is difficult to estimate and measure. Even for software reliability, which is the most popular measurement for software quality, there are few valid prediction or measurement tools and techniques available. Most tools estimate software reliability only after testing has begun, and are therefore of little use to analyzing software quality during the analysis and design phases of a software effort. Even those tools that are useful after testing are often theoretical and unproven. Measurement of quality is also a problem. There is currently no standard list of quality metrics nor any established methods for collecting data for quality metrics.

SOFTWARE QUALITY IMPROVEMENT

The lack of validated software TQM prediction or measurement methodologies should not discourage consideration of software TQM. There are numerous activities which will result in higher quality and improved reliability for computer software, although the results can not always be estimated or measured. Use of higher- order programming languages, such as the DoD-Standard Ada language, has been shown experimentally to reduce software errors and improve overall software quality. Use of modern design practices such as structured design and coding has shown similar results.

Independent evaluation of the software throughout the development process also improves quality. This evaluation can range from internal informal walk-throughs to the employment of an independent verification and validation contractor to continuously evaluate a developer's procedures and products. These are merely examples of the myriad of activities that, when used, can greatly improve the quality of computer software.

Software TQM currently lags that of other system elements both in emphasis and in the capability for estimation and measurement. However, there are many activities that can result in higher quality software. Employment of these activities, and serious consideration of software TQM in general, is essential for modern systems which are usually very software-intensive.

4.7   Metrics

DEFINITION OF METRICS

A chapter on quality would not be complete without an discussion of metrics, which may be defined in broad terms as meaningful measures of quality, performance, or cost effectiveness. For these measures to be meaningful, however, they must present data that allow us to take appropriate improvement actions. They should also be customer oriented and should foster process understanding, thereby motivating actions to continually improve the way we do business. In this way, metrics can support organizational strategic planning by allowing us to get insight into how, as in our case, the acquisition processes are meeting user needs. It should, however, be pointed out that measurements by themselves will not always result in process improvements, but meaningful metrics will.

Metrics also play an integral role in linking organizational processes to the achievement of an organizational plan. Metrics are the means by which we gauge process toward meeting corporate goals and objectives. As shown in Fig. 4.2, corporate planning starts with a statement of the organizational mission and its vision for the future. A strategic analysis of major internal and external influences identifies key issues which must be addressed to achieve the long-term vision. The analysis and key issues define where current capabilities and future organizational requirements are not aligned. A strategic plan culminates in the identification of goals which specifically address these issues.

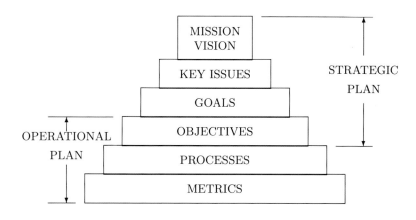

Fig. 4.2: Corporate planning framework.

Goals and their subordinate objectives chart the appropriate course of action. Those objectives, in many cases, include the improvement of processes and tasks which involve local organizations and personnel. At this level we must align selected metrics to organizational objectives and identify which processes are targeted for improvement through their application. Organizational objectives and processes, and the metrics which support them, make up the organization's operational plan. One word of caution may be necessary while introducing metrics at the organizational level. Metrics should never be used as a gauge for evaluating personnel individual performance. Hence an important aspect of a good metric is that it is not perceived as threatening to the personnel. Good metrics will gauge progress in reaching the processes' desired outcomes. This will lead to accomplishment of organizational objectives, and consequently result in achieving goals and visions.

A succinct definition of metrics was used by Lieutenant General Thomas R. Ferguson, Jr., former Commander of the Aeronautical Systems Center, Wright-Patterson AFB, in his policy letter:

> Measurement is a fundamental part of good management. Metrics are invaluable from both a program management and a process improvement perspective. As such, we need metrics to be an integral part of daily operations throughout Air Force Systems Command.* Measuring processes provides the basis for appropriate action(s) to identify opportunities for constructive changes and continuous process improvement. Metrics allow us to baseline where we are, identify the impediments to the process, and track the impact of management actions on processes and other process changes.

USE OF METRICS

Metrics must represent measures useful to their users. To accomplish this a good metric should have the following characteristics:[16]

1. It is accepted as meaningful to the customer.

2. It tells how well organizational goals and objectives are being met through processes and tasks.

3. It is simple, understandable, logical, and repeatable.

4. It shows a trend.

5. It is unambiguously defined.

6. Its data is economical to collect.

7. It is timely.

8. It drives the "appropriate action."

---

*Now merged into the new Air Force Materiel Command.

Developing good metrics is a challenge to every organization in the world of acquisition management of weapon systems. Good metrics will always lead to process improvements, and only those organizations with well-defined improvements process can expect to survive in today's turbulent environment – an environment of constant change.

Metrics are as good as the data which supports them. Measuring a product of a process may not be enough. Measuring the process is more beneficial, but oftentimes that process is not consistent across many organizations, much less many programs. For example, data and graphs alone may not be enough to effect the proper constructive changes for the acquisition process. It may be necessary to "peel the onion" through several lower level metrics before solid actions can be taken based on the available data. Some metrics even measure multiple processes and may simple point you in the direction of another, more specific, measure. Metric have their faults and shortcomings, but process improvement can never occur unless you start measuring something. After gathering data, you will naturally change your metric. You may even delete the metric altogether and/or add a new one. Perhaps the metric's original design was inadequate and did not capture the process. Maybe the metric cannot support any process improvement. Or maybe the metric has given you insightful data which is then refined to give you a closer look at a particular portion of the process. Whatever the case, you have to start somewhere.

METRICS TOOLS

The same statistical tools which support effective process control can be useful in developing good metrics. The tools most commonly mentioned in TQM literature are the following:[17]

- Control charts

- Run charts

- Flow charts

- Cause and effect (or fishbone) diagrams

- Histograms

- Pareto diagrams

- Scatter diagrams

- Check sheets

ACQUISITION PROGRAM TRACKING SYSTEM (APTS)

An excellent example of the application of metrics is the metrics system developed by the Aeronautical Systems Center (ASC), Air Force Materiel Command (AFMC), at Wright-Patterson Air Force Base, Ohio.[18] This system is an automated data base and is referred to as the Acquisition Program Tracking System (APTS). It automates the collection and analysis of all acquisition metrics. Essentially, it is an Executive Information System designed to track all organizational metrics. The data base contains information related to organizational objectives, including responsiveness, cost schedule, performance, user engagement, acquisition excellence, work environment, and technology transition. All information in the APTS data base is attached to a particular ASC organization and/or program , and data charts ASC's progress towards meeting its objectives in accordance with the AFMC goals. Most importantly, APTS is a source of data which drives the "appropriate actions" toward continual improvements. APTS can drive "appropriate actions" to help reverse negative trends. Similarly, APTS can promote positive trends throughout the organization through the sharing of successful ideas (benchmarking).

ASC metrics are broken down into two broad categories: program metrics and organization metrics. Program metrics pertain to specific, baselined, acquisition programs in support of external customers, typically operational commands. Organization metrics pertain to processes and projects within organizations and offer insight to internal efficiency and effectiveness. For example, a single organization may report metric data on several different acquisition programs, but will also report consolidated, unit-wide organizational metric data.

APTS is a management tool. It empowers the people to make informed decisions on processes measured in ASC. Information from APTS can be powerful. The direct link between APTS and the Acquisition Program Baseline (APB) makes tracking to a baseline simple and straightforward. Similarly, lower level, non-baselined data on ASC programs or projects may be tracked to give program/project manager direct insight into the program's or project's performance according to plan. APTS data allow us to take appropriate actions before the need for oversight. APTS is another management tool to help do the job more effectively, or identify impediments to the process which can be addressed by upper-level management.

References

[1] Deming, W.E, Improvement of Quality and Productivity Through Action by Management, National Productivity Review, Winter 1981-82.

[2] Deming, W.E., *Out of the Crisis*, MIT Center for Advanced Engineering Study, Cambridge, MA, 1986.

[3] Roadmap for Change/The Deming Approach, videotape, Encyclopedia Britannica Educational Corporation, Chicago, IL, 1983.

[4] Juran, J.M., *Juran on Planning for Quality*, Free Press, New York, and Collier Macmillan, London, 1988

[5] Juran, J.M., *Juran on Leadership for Quality*, Free Press, New York, and Collier Macmillan, London, 1989

[6] Juran, J.M., *Juran on Quality by Design*, Free Press, New York, and Collier Macmillan, London, 1992

[7] Taguchi, Shin and D.M. Byrne, "The Taguchi Approach to Parametric Design," Quality Progress, pp. 19-26, 1987

[8] Ross, Phillip J., *Taguchi Techniques for Quality Engineering*, Mc-Graw-Hill, New York, 1988

[9] Crosby, P.B., *Quality is Free: the Art of Making Quality Certain*, McGraw-Hill, New York, NY, 1979.

[7] Unal, Resit, "Parametric Analysis of Quality Costs," Proceedings of the International Society of Parametric Analysis, 12th Annual Conference, pp. 91-102, 1990.

[11] Taylor, F.W., *Scientific Management*, Harper, 1947.

[12] Scherkenbach, W.W., *The Deming Route to Quality and Productivity: Road Maps and Roadblocks*, CEE Press, George Washington University, Washington, DC, 1987.

[13] Walton, M., *The Deming Management Method*, Putnam Publishing Group, New York, NY, 1986.

[14]Juran, J.M., Ed., *Juran's Quality Control Handbook*, 4th Edition, McGraw-Hill, New York, 1988.

[15]Peters, T.J., *Excellence in the Public Sector*, Videorecording, Enterprize Media, Inc., Boston, MA, 1989.

[16]"The Metrics Handbook," Air Force Systems Command, Andrews Air Force Base, MD, August 1991, p. 2-1.

[17] Ibid., pp. A-1 – A-38.

[18]"ASD Metrics Handbook," Headquarters Aeronautical Systems Division, Wright-Patterson AFB, OH, ASD Pamphlet 700-8, 30 April 1992, pp. 21-30.

Chapter 5

INTEGRATED WEAPON SYSTEM MANAGEMENT

5.1 Introduction

In 1992, the Air Force created a new command, the Air Force Materiel Command (AFMC), by merging the Air Force Systems Command (AFSC) with the Air Force Logistics Command (AFLC). AFSC was responsible for the development, acquisition, and delivery of weapon systems, as well as Research and Development programs to ensure technological superiority, while AFLC was responsible for ensuring readiness and sustainability of Air Force systems worldwide. This merger embraced the new philosophy of Integrated Weapon System Management (IWSM) as the guiding management concept for the new command. IWSM attracted a lot of attention in DoD and both the Army and Navy are considering similar management concepts. The AFMC's projected role was perhaps best described by AFSC and AFLC commanders, as quoted below:

> In a world characterized by tremendous economic, political, social and technological change, our national interests are evolving. The Air Force is undergoing the most intensive period of change since it became a separate service 44 years ago. In his white paper "Global Reach...Global Power", the Secretary of the Air Force defined our primary operational missions and described a new, emerging organizational structure to execute those missions. Air Force units are being asked to curtail management layers, speed decision making, reduce overhead, streamline operations and devise improved business practices.

> Air Force Materiel Command's role is to turn global power and reach concepts into capabilities–to design, develop and support the world's best air and space weapons systems. The cornerstone of Air Force Materiel Command will be integrated weapon

> system management–the "cradle to grave" management of all
> Air Force Systems. This approach provides a single focal point
> for our customers–that single focal point will be the system pro-
> gram director, who will have responsibility for all aspects of a
> system or commodity throughout its life. It increases the system
> program director's authority and flexibility, integrates all crit-
> ical processes and eliminates the "seams" that currently exist
> between development and support.*

Over the last four decades, Air Force Systems Command and Air Force
Logistics Command pursued textbook concepts of product management and
organizational design. Each command optimized its strategies toward the
assigned mission. Air Force Systems Command focused on the front end of
the weapons system life cycle and stressed the technology and acquisition
elements, while Air Force Logistics Command focused on wartime readiness
and sustainability for the long haul. Bridging organizations were often
established to cross the "seams" created along mission boundaries.

Desert Storm was a dramatic test of our progress and prowess. The
logistics system employed in Desert Storm demonstrated its excellence by
moving massive amounts of equipment and supplies across the globe and
sustaining our forces in battle. Likewise, our weapon systems proved, beyond a doubt, their pre-eminent technological position in the world. However, as in the civilian sector, our national military organizations had to
change to meet the dramatically altered international environment and the
challenges of tomorrow.

In 1987, Air Force Systems Command and Air Force Logistics Command began a new quest for total quality leadership. Their customers were
demanding it and the Air Force had many ideas on how to improve their
processes. Under the Defense Management Review, these commands continued streamlining and integrating many elements of their business, moving closer to a single, uniform acquisition and support process for the total
life cycle. As their organizations transitioned and world events unfolded, it
became clear that the projected smaller, more flexible and responsive Air
Force could be better served by a single command charged with supporting
all the equipment needs of the warfighting commands.

Creating a new command and integrating all the elements of weapon
system management allowed the Air Force to better respond to the needs
of the warfighting commands. Air Force can continue to develop better
business practices as the distinctions and ownership boundaries between

---

*Statement by General Ronald W. Yates, Commander, Air Force Systems Command,
and General Charles C. McDonald, Commander, Air Force Logistics Command, White
Paper on IWSM, AFSC and AFLC, 28 January 1992

product development and support are eliminated. The AFMC's job is now to plan with its customers, explore their operational missions and tasks, and find ways to give them what they need, when and where they need it.

5.2   Vision of the Future[1]

The Secretary of the Air Force and his senior leaders captured the vision of the future – "Force people building the world's most respected air and space force – global power and reach for America." Air Force Materiel Command is a full partner on the Air Force team in every respect. It supports this corporate vision by providing the technologies, systems, products and support its customers require in peace and war.

Air Force Materiel Command's operational philosophy is to support the integrated weapon system management concept, based on three simple components:

- "Cradle-to-Grave."

- Single Face to the User.

- Seamless Organization.

"Cradle to Grave:" The scope of this management task must start at the beginning of a system or commodity and continue through its complete life cycle–"cradle to grave", from milestone I (demonstration/validation phase) through system retirement/cancellation. This insures that management considers the impact of its decisions not only on development activities, but also on the operational phase, which can span several decades.

Single Face to the User: There should be only a single face to the user, simplifying the customer-supplier relationships. A single organization, the system program office, manages the weapon system or commodity. This organization is headed by a single individual: the System Program Director (SPD), Product Group Manager (PGM), or Materiel Group Manager (MGM). The make-up of the organization will change over time and elements will likely be at multiple locations. However, program management responsibility never leaves the hands of the single manager.

Seamless Organization: The weapon system or commodity program office is a seamless organization, operated with critical processes that are integrated across the life cycle. Other functions, such as test centers, laboratories, and depot maintenance, will provide the required support to the program offices.

These concepts are revolutionary; they are a step toward the future and fundamentally change current business practices. They demand cultural changes of every member of the Air Force Materiel Command.

5.3   Emerging Principles[1]

The IWSM operating concept is based on a set of seven guiding principles listed below:

- Increasing Single Manager Authority.

- Creating Single Business Decision Authority.

- Inserting Technology.

- Creating Integrated Product Development Teams.

- Maintaining Management Continuity.

- Building New Partnerships.

- Consolidating Air Force Acquisition.

Increasing Single Manager Authority: The person who is in charge of the system program office, the single manager or director, has increased authority over a wider range of decisions and resources. The decisions which the director receives from the Air Force Acquisition Executive come through a streamlined chain of command with only one intermediate layer– the Program Executive Officer or designated acquisition commander. To empower the Integrated Weapon System Management concept, program direction from the acquisition executive is being consolidated to cover the total life cycle. Resource decisions already made in this area include: converting interim contractor support funds from operations and maintenance to procurement accounts; sustaining engineering funds are being shifted from operations and maintenance to research and development accounts; and, initial common support equipment funds are being transferred to weapon system procurement accounts. This new, expanded resource responsibility allows the system program director to consider new and innovative ways to develop and support the weapon system.

Creating Single Business Decision Authority: With a seamless organization and integrated processes across the life cycle, the single manager is the single business decision authority. Program management decisions that were divided across the seams of the commands are consolidated in the system program office. The director is responsible for all systems engineering decisions. Typical of the new approach is the Low Altitude Navigation and Targeting Infrared System for Night (LANTIRN) program which now has a single configuration control board, integrated technical interchange meetings, and common operating instruction for engineering change proposals, modifications, deviations and waivers. The consolidated configuration control boards include depot modifications and interfaces with the commodity groups. Appropriate representatives from across Air Force Materiel Command are on this board, and the system program director makes the decisions.

Contracting is a very important part of the single business decision authority of the system program director. The director has a dollar-threshold at which a contract must be referred to command headquarters for review and approval. Current thresholds are not the same across the commands' field organizations. For example, approval levels for like items being purchased for the F-15 differed by $75 million between the product and logistics centers. Air Force Materiel Command uses now one set of consistent thresholds, based on the type of product or service to be purchased. Most review and approval thresholds are set at significantly higher dollar amounts. This achieves consistency in business practices and provides authority for most contracting decisions to the system program directors.

Inserting Technology: The IWSM concept created a new technology insertion process where none previously existed. The creators, developers and users of technology now play key roles in drafting a technology insertion investment plan. They do this in concert with operating commands' requirements planners as well as development planners and science and technology managers in Air Force Materiel Command. They use this plan to coordinate program management direction from the acquisition executive and develop appropriate infrastructure direction to field activities. Thus the single managers are able to take appropriate actions for inserting technology into new systems, fielded systems and commodities. The center commanders make technology insertion decisions regarding the command's infrastructure and the new technology insertion process is used to better focus new technology into existing weapons systems, improve the competitive position of our depots, and redress our environmental problems.

Creating Integrated Product Development Teams: The Single manager for a new weapon system is located at a product center. The system program office, formed with the support and resources of the product enter commander, will use a management philosophy known as Integrated Product Development. Many of the pilot programs, F-15, F-16, B-2, F-22 and others, found that integrated product development teams are essential to make the new core processes work most effectively. Teams are formed with product emphasis and are made up of designers, configuration managers, manufacturing experts, logisticians, testers and supported by contracting officers, financial managers and other disciplines as required. A system support manager from a logistics center is assigned to the program office and reports to the System Program Director. System program directors for commodities are located at logistics centers, have a product focus, and call upon the product centers for assistance in commodity development programs. This emphasis on product development reinforces the "cradle-to-grave" concept so important to providing the user with never-ending support. It also speeds up the development cycle by enabling the teams to solve problems in parallel, rather than passing it from design to engineering to manufacturing to logistics support.

Maintaining Management Continuity: A weapons System Program Office (SPO) remains at the product center until weapon system development is complete. The office may relocate to a logistics center later in its life when the predominant activity is operational support. The emphasis is on management continuity, not rapid transfer. This permits the centers to concentrate on what they do best. The product enters are focused on converting systems requirements into operational systems. The logistics centers strength is supporting weapon systems in the field, providing the critical elements of combat readiness and sustainability that warfighters need to employ the systems. Program office location will build on and sustain these strengths throughout the life cycle. Following this principle, the B-1 bomber and silo-based intercontinental ballistic missile offices at product centers were closed and relocated at logistics centers. Because they face modifications with major development activity, lead management responsibilities for the Airborne Warning and Control System program and F-16A/B fighter program were relocated from logistics centers to product centers.

Building New Partnerships: Integrated weapon system management permits new teamwork and partnerships to develop in Air Force Materiel Command. As the product and logistics centers focus on their areas of expertise, the single manager can go anywhere in the command to get the needed help. The single manager can anticipate that some of the expertise he needs will not be co-located in the program office. For example, the director will look to the system support manager, who may be located at a logistics center, to carry out the assigned tasking by coordinating across all other centers. The Joint Surveillance and Attack Radar System program recently demonstrated its team work as engineers from the logistics center supported the development of specifications for the program's integrated support facility. These partnerships are especially necessary for commodity management, which is currently done at both product and logistics centers. The product centers rely on the centers of management and technical expertise fund in the commodity groups at logistics centers, rather than build or maintain this capability internally. Conversely, the System Program Director at a logistics center uses a Development System Manager at a product center when a major development activity is assigned within the program, rather than creating an internal capability.

Consolidating Air Force Acquisition: It is the objective of the Air Force to concentrate all acquisition in Air Force Materiel Command. Accordingly, the Air Force Chief of Staff and the Air Force Acquisition Executive requested a review of communication/computer system acquisition and support in Air Force Communications Command. To test the integrated weapon system management process in the communication/computer area, pilot programs were selected and added to the existing Air Force Materiel Command pilot programs. This test resulted in the decision to apply integrated weapon system management principles to all communications/computer systems acquisition and support programs in Air Force Communications Command, and to transfer these functions to Air Force Materiel Command on 1 July 1993.

References

[1] Air Force Materiel Command, "Integrated Weapon System Management," Pamphlet 800-60, 31 March 1993.

Chapter 6

SCIENCE AND TECHNOLOGY MANAGEMENT

6.1 Introduction

The United States has a primary national security objective to preserve
the U.S. as a free nation with its fundamental institutions intact. Militar-
ily, this translates into sustaining a superior fighting force to deter attacks
against the U.S., our allies, and vital interests worldwide; to prevent an
enemy from politically coercing the U.S.; and if deterrence fails, to fight at
a level of intensity and duration necessary to achieve U.S. national security
objectives. One strategy is to develop systems that are superior to those of
our adversaries, but in lesser quantity than theirs. The emphasis of quality
over quantity as an operational utility has been adopted and the acquisition system modified to develop superior technology and to transition that
technology rapidly into operational systems.

With this strategy, science and technology is vital to the U.S. posture
of deterrence of wars and the stability in the world. The Science and
Technology (S&T) programs in the three services, with their trained and
experienced laboratory personnel and state-of-the-art facilities, exploit scientific breakthroughs and develop a credible array of technologically superior options for weapon systems to support any defense posture. They
also contribute to the foundation of the industrial base and improve the
global competitiveness of commercial products that benefit from the defense R&D. To sustain the technological and industrial superiority, it is
important that we transition our scientific breakthroughs and technological
options as quickly as possible from our laboratory benches to our industrial
production floors.

In order to insure technological superiority and meet national defense
needs, a sound technology investment strategy must be formulated, the
right technology must be developed, and a plan for technology transition
must be executed. Expeditiously transitioning the right technology into our

future weapon systems requires regular interaction with operational users as well as teamwork across product divisions, program offices, development planning and engineering organizations, and the laboratories. During this transition period, the field organizations must understand what their roles are and how to foster efficient use of S&T resources.

S&T activities include three budget categories: Basic Research (Budget Category 6.1), Exploratory Development (Budget Category 6.2), and Advanced Technology Development (Budget Category 6.3A), shown in Fig. 6.1.

Basic Research is the fundamental investigation of new ideas and scientific principles. Extensive research is conducted through universities and research centers to establish the foundation for further development.

Exploratory Development is the assessment of scientific and technological advances warranting further examination. Those technologies are pursued for feasibility to future military application. This activity is managed by the laboratories and involves both in-house investigation and contracts with industry and academia.

Advanced Technology Development is the risk reduction demonstration of promising technologies and involves the manufacture of one-of-a-kind components, subsystems, and systems. The technologies developed under this budget category are defined as either critical experiments or advanced technology demonstrations (ATDs); the Air Force designates ATDs as the advanced technology transition demonstrations (ATTDs). A critical experiment is a laboratory project in which the technical feasibility is demonstrated at the component or subsystem level, generally in a laboratory environment. Critical experiments may lead to subsequent ATDs. ATDs are laboratory projects with specific objectives of meeting users' defined needs through risk reducing "proof of principle" demonstrations conducted at the subsystem or higher level in an operationally realistic environment. Laboratories manage advanced technology development mainly through contracts with industry. It is important at this stage in technology development that industry accept the technology and the level of risk so that they can propose these technologies to system program offices for further weapon system development.

These activities support the overall systems development program which includes the demonstration and validation of systems, engineering and manufacturing development (EMD), and management and support (Budget Categories 6.3B, 6.4, and 6.5), as shown in Fig. 6.1.

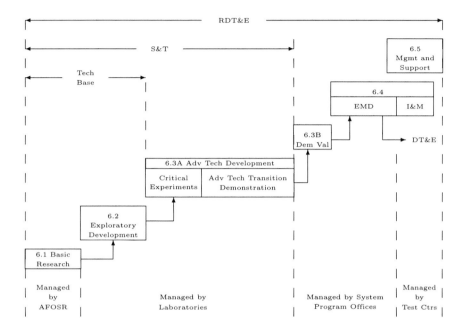

Fig. 6.1: Budget Categories 6.1 through 6.5 relationships.

6.2   DoD Science and Technology Program[1,2]

In 1992, a new Science and Technology strategy was formulated. This strategy is based upon seven thrusts which are oriented toward significant improvements in war fighting capability.[3] The core of this strategy is to :

- Provide for the early, intensive, and continued involvement of warfighters.

- Fuel and exploit the information technology explosion.

- Conduct extensive, and realistic advanced technology demonstrations (ATDs).

These demonstrations range from assessing the military utility of new technological concepts in the laboratory to integrating and evaluating technology in as realistic operational environment as possible. Technology demonstrations are nothing new. The Have Blue aircraft demonstrated that stealth was feasible in flying aircraft prior to the development of the F-117.

Assault Breaker demonstrated the radar technology for the Joint Surveillance and Target Acquisition Radar System (JSTARS) and the Advanced Tactical Missile (ATACM) system. The Milimeterwave and Microwave Monolithic Integrated Circuits (MIMIC) program has demonstrated the ability to produce low-cost integrated circuits.

The Director of Defense Research and Engineering (DDR&E) takes the lead in the development of, and has decision authority over, all major aspects of the Science and Technology program in the Department of Defense. DDR&E directs the Secretaries of the Military Departments and heads of other components of DoD, when necessary, with respect to all activities supported by funds from budget categories 6.1, 6.2, and 6.3A; DDR&E also exercises oversight of and provides support for the technological aspects of the Strategic Defense Initiative (SDI). Thus, while the actual program strategies are executed by the military departments and defense agencies, they are carried out under the guidance and direction of the DDR&E.

The Science and Technology strategy is imbedded in and aligned with the President's National Security Strategy , U.S. National Military Strategy, and the Defense Department's revised acquisitions strategy. One of the eight strategic principles* articulated in the *National Military Strategy (1992)*,[2] is Technological Superiority that capitalizes on our enduring strengths and allows us to exploit the weaknesses of our potential adversaries. The United States must rely on a strong research and development establishment to provide the systems that reduce the risks to our forces and those of our allies and enhance the potential for the swift, decisive, and economical termination of conflict, with minimal casualties, on favorable terms.

In peace, technological superiority is one of the most important elements of deterrence. In crisis, it provides a wide spectrum of options to the Commanders-in-Chief (CINCs) and the National Command Authorities. In war, it enhances combat effectiveness and reduces the loss of personnel and equipment. For example, the Gulf War demonstrated the benefits of a superior intelligence capability and superior weapons and supporting systems, all used by highly trained professional men and women, from the soldier to the CINC. Therefore, advancement and protection of technology must be a national security obligation.

---

*The eight strategic principles are: Readiness, Collective Security, Arms Control, Maritime and Aerospace Superiority, Strategic Agility, Power Projection, Technological Superiority, and Decisive Force.

TECHNOLOGY THRUSTS

The seven Science and Technology (S&T) thrusts identified in the 1992 Defense S&T Strategy document[3] represent the DoD assessment of the demands being placed on the S&T program by the users' most pressing military and operational requirements. As national security requirements, operational needs, and technology evolve, additional thrusts can be added and existing thrusts can be modified. Thus the planning process must be flexible to accommodate the changing needs. The seven 1992 thrusts are summarized below:

1. Global Surveillance and Communications: The ability to project power requires a global surveillance and communications capability that can focus on a trouble spot, surge in capacity when needed, and respond to the needs of the commander.

2. Precision Strike: The desire for reduced casualties, economy of force, and fewer weapons platforms demands that we locate high-value, time-sensitive fixed and mobile targets and destroy them with a high degree of confidence within tactically useful timelines.

3. Air Superiority and Defense: The need to defend deployed military forces, and help defend allies and coalition partners, from the growing threat of high performance aircraft and ballistic and cruise missiles, and the need to maintain decisive capabilities in air combat, interdiction, and close air support, require a strong effort in missile defense and air superiority.

4. Sea Control and Undersea Superiority: The ability to maintain overseas presence, conduct forcible entry and naval interdiction operations, and operate in littoral zones, while keeping loses to a minimum, presupposes a strong capability in sea control and undersea warfare.

5. Advanced Land Combat: The ability to rapidly deploy our ground forces to a region, exercise a high degree of tactical mobility, and overwhelm the enemy quickly and with minimal casualties in the presence of a heavy armored threat and smart weaponry requires highly capable land combat systems.

6. Synthetic Environments: A broad range of information and human interaction technologies must be developed to synthesize present and future battlefields. We must therefore synthesize factory-to-battlefield environments with a mix of real and simulated objects and make them accessible from widely dispersed locations. Integrated teams of users, developers, and/or testers will be able to interact effectively. Synthetic environments will prepare our leaders and forces for war.

7. *Technology for Affordability:* Technologies that reduce unit and life cycle costs are essential to achieving significant performance and afford-ability improvements. Manufacturing process and product performance issues are integral parts of the program. Advances are particularly needed in technologies to support integrated product and process design, flexible manufacturing systems that decouple cost from volume, enterprise-wide in-formation systems that improve program control and reduce overhead costs, and integrated software engineering environments.

KEY TECHNOLOGY AREAS

In addition to the technology thrusts, the Department of Defense identi-fied in 1992 eleven Key Technology Areas, which are essential for achieving the goals and objectives of the technology thrusts. Each Key Technology Area supports one or more Technology Thrusts. The DoD Key Technol-ogy Plan[4] provides technology development roadmaps for the development and maturation of the technologies needed to achieve the stated goals of the thrusts. The plan further provides for the investigation of innovative technologies that could have a significant impact on military performance across a broad spectrum of applications. The following is a summary of the 1992 Key Technology Areas:

1. *Computers:* High performance computing systems (and their soft-ware operating systems) that provide orders-of-magnitude improvements in computational and communications capabilities as a result of improve-ments in hardware, architectural designs, networking, and computational methods.

2. *Software:* The tools and techniques that facilitate the timely genera-tion, maintenance, and enhancement of affordable and reliable applications software, including software for distributed systems, database software, ar-tificial intelligence, and neural nets.

3. *Sensors:* Active sensors (with emitters, such as radar and sonar), passive ("silent") sensors (e.g., thermal imagers, low light level TV, and infrared search and track systems), and the associated signal and image processing.

4. Communications Networking: The timely, reliable, and secure production and worldwide dissemination of information, using shared communications media and common hardware and applications software from originators to DoD consumers, in support of joint-service mission planning, simulation, rehearsal, and execution.

5. Electronic Devices: Ultra-small (nano-scale) electronic and optoelectronic devices, combined with electronic packaging and photonics, for high speed computers, data storage modules, communications systems, advanced sensors, signal processing, radar, imaging systems, and automatic control.

6. Environmental Effects: The study, modeling, and simulation of atmospheric oceanic, terrestrial, and space environmental effects, both natural and man-made, including the interaction of a weapon system with its operating medium and man-produced phenomena, such as obscurants found on the battlefield.

7. Materials and Processes: Development of man-made materials (e.g., composites, electronic and photonic materials, smart materials) for improved structures, higher temperature engines, signature reduction, and electronics, and the synthesis and processing required for their application.

8. Energy Storage: The safe, compact storage of electrical or chemical energy, including energetic materials for military systems.

9. Propulsion and Energy Conversion: The efficient conversion of stored energy into usable forms, as in fuel efficient aircraft turbine engines, and hypersonic systems.

10. Design Automation: Computer-aided design, concurrent engineering, simulation, and modeling, including the computational aspects of fluid dynamics, electromagnetics, advanced structures, structural dynamics, and other automated design processes.

11. Human-System Interfaces: The machine integration and interpretation of data and its presentation in a form convenient to the human operator; displays; human intelligence emulated in computational devices; and simulation and synthetic environments.

RELATIONSHIP TO CRITICAL TECHNOLOGIES

The 1992 DoD Key Technologies Plan was intended to fulfil the requirements of PL 101-189, National Defense Authorization Act for Fiscal Years 1990 and 1991, as amended by PL 101-510, and serve as the 1992 edition of the Defense Critical Technologies Plan. The 20 Defense Critical Technologies identified by DoD in 1990, and the 21 technologies identified in 1991, were selected through a much less focused process. On the other hand, the 1992 list of key technologies was selected because of their importance to achieving the goals of the S&T Strategy Thrusts. The 1991 list essentially differed from the 1990 list only by the addition of the Flexible Manufacturing Technology. The 20 technologies from the 1990 plan were discussed at length in an Air Force Institute of Technology textbook, which provided for each technology a description of the physical and engineering principles involved, technology description, and impact on future weapon systems.[5] This text was developed with the specific objective of providing information on the direction of future science and technology research and education in response to the projected critical technology needs for national defense.

There is considerable similarity between the 1991 Critical Technologies and the 1992 Key Technologies lists. Table 6.1 presents the relationship between the two taxonomies. It shows the extent to which the Critical Technologies map into the Key Technologies, not the other way around. The numbers in the table reflect the extent to which Critical Technology is covered by a particular Key Technology on a scale 1 to 10. The highest number (10), indicated in the table by an asterisk (*), means that there is an almost perfect one-to-one mapping. The table shows that most of the Critical Technologies are in fact in one or more Key Technology Areas.

6.3 Air Force Management of Science and Technology

In what follows, the specifics of Air Force management of science and technology will be described. The S&T program is the innovative arm of the Air Force reaching into the future. In FY 92, the Air Force S&T budget was approximately $1.7 billion, executed by four interdisciplinary laboratories that operate $3.0 billion of world-class facilities. Direction and control of the Air Force S&T program are exercised by the Air Force Acquisition Executive (AFAE) through the Technology Executive Officer (TEO) to the laboratory commanders and directors (see Fig. 6.2). The AFAE relies on the TEO to establish management control of the Air Force S&T program to assure the best investment of the Air Force technology money. Each laboratory has clearly defined areas of technical responsibility, and in conjunction with its parent product division plans the overall transition of technology

Table 6.1: Relationship Between Critical and Key Technologies

	CRITICAL TECHNOLOGIES	KEY TECHNOLOGIES										
		1	2	3	4	5	6	7	8	9	10	11
1	Semiconductor Materials and Microelectronic Circuits					*						
2	Software Engineering		*									
3	High Performance Computing	*										
4	Machine Intelligence and Robotics		2	2	2						2	2
5	Simulation and Modeling										5	5
6	Photonics			2		3						
7	Sensitive Radar			*								
8	Passive Sensors			*								
9	Signal and Image Processing			*								
10	Signature Control			1			1	4			4	
11	Weapon System Environment						*					
12	Data Fusion											*
13	Computational Fluid Dynamics										*	
14	Air-Breathing Propulsion									*		
15	Pulsed Power								*			
16	Hypervelocity Projectiles and Propulsion								3	3		
17	High Energy Materials								*			
18	Composite Materials							*				
19	Superconductivity	1		2		4		1	1	1		
20	Biotechnology			2				2				
21	Flexible Manufacturing											*
	KEY TECHNOLOGIES	1	2	3	4	5	6	7	8	9	10	11
1	Computers											
2	Software											
3	Sensors											
4	Communications Networking											
5	Electronic Devices											
6	Environmental Effects											
7	Materials and Processes											
8	Energy Storage											
9	Propulsion and Energy Conversion											
10	Design Automation											
11	Human Systems Interface											

to Air Force systems. Recently, the Air Force reorganized the laboratory structure into four major laboratories as indicated below:

- The Armstrong Laboratory (Human Systems Center (HSC)), Brooks AFB TX, responsible for human systems technology;

- The Phillips Laboratory (Space and Missile Systems Center (SMSC)), Kirtland AFB, NM, responsible for space and missile technology;

- The Rome Laboratory (Electronic Systems Center (ESC)), Griffiss AFB NY, responsible for command and control, communications, and intelligence technology; and

- The Wright Laboratory (Aeronautical Systems Center (ASC)), Wright-Patterson AFB OH, responsible for air vehicles and air delivered weapons.

In addition, Air Force Materiel Command (AFMC) has a Memorandum of Understanding with the Civil Engineering Laboratory of the Air Force Civil Engineering Support Agency (AFCESA) to develop technologies for civil engineering and environmental quality.

6.4   Program Development

The Air Force S&T program is designed to be responsive to user requirements and topdown guidance, and be able to rapidly exploit innovation which was unforeseen at the time the requirements and guidance were prepared.

A vast number of component level technologies are usually advanced in order to create technically superior weapon systems. Literally hundreds of individual technical advancements are needed to advance the state-of-the-art. Every year, in response to DoD, AFAE, and TEO strategic planning guidance, the laboratories document the details of their technology development strategy in 12 technology area plans (TAPs) for approval. These TAPs cover the areas of aeropropulsion and power, air vehicles, avionics, conventional armament, advanced weapons, space and missiles, human systems, C3I, civil engineering and environmental quality, geophysics, materials, and basic research. The Basic Research TAP is prepared by the Air Force Office of Scientific Research (AFOSR) and provides the single integrated plan for the 6.1 research investment. The other 11 TAPs cover the Air Force's 6.2 Exploratory Development and the 6.3A Advanced Technology Development efforts.

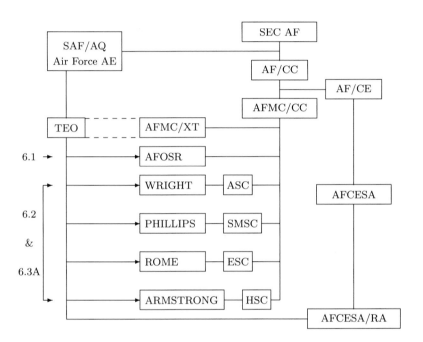

Fig. 6.2: Air Force S&T management.

6.5 Program Evaluation[6]

The vast scope of work must be evaluated by the TEO (1) for compliance with DoD and AFAE guidance so that the AFAE can approve the program for execution, and (2) for quality and relevance in order to continuously improve the program. The TEO analysis of the program involves a diverse measurement and evaluation process. Using feedback from the Air Force Scientific Advisory Board (SAB) (for technical quality), the using commands (for operational relevance of ATTDs), and the development planners (for relevance of the technology areas), the TEO keeps a balanced focus on customer needs while advancing technology innovation in new research areas. The SAB reviews the S&T program for technical innovation, uniqueness, past successes, and peer assessment. The major air commands (MAJCOMs), who are the ultimate customers for the Air Force's S&T investment, annually score the ATTDs being demonstrated by the laboratories. Since the ATTDs are the last step in the laboratory development prior to incorporation of the technology in an AF weapon system, the horizon for utility for the MAJCOM is near enough to render a legitimate judgement of the importance of the ATTDs to that MAJCOM.

The major thrusts of the 11 TAPs containing the 6.2 and 6.3A programs are evaluated annually by the development planning organizations of the Air Force product divisions. The product divisions contain the System Program Offices that develop and manage the production of the Air Force's weapon systems. Each product division has a developmental planning organization that works with the war-fighting commands to create the concepts for the next generation of Air Force weapon systems. Hence, the development planners are good judges of the potential benefits of emerging technology to the performance and life cycle costs of future systems. The 6.1 TAP is evaluated by the laboratory commanders and directors with the advice and assistance of their chief scientists.

6.6   TEO Analysis

Based on the SAB assessment of program quality and technical innovation, and the customers assessment of program importance to AF capability needs, the TEO and his staff conduct an analysis of the S&T program. The TEO's objective is to develop recommendations for the AFAE that will continuously improve both the quality and relevance of the S&T program's content. The entire closed-loop process of guidance, development, evaluation, analysis, and feedback is depicted in Fig. 6.3. In executing this process, the TEO must give particular attention to the protection of technical innovation which has not matured sufficiently to achieve customer recognition.

The feedback control process offers a very efficient and effective alternative to the laborious strategic planning process that would be necessary to individually authenticate the more than 5,000 work units of the Air Force S&T program. It provides the macro view of the S&T program that the AFAE needs to be confident that the Air Force laboratories are "doing the right technology."

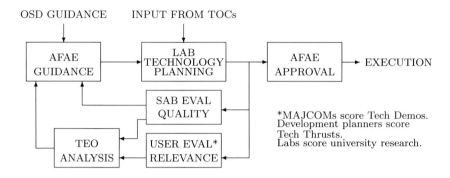

Fig. 6.3: Air Force S&T investment planning process.

6.7 Technology Transition Planning

Doing the right technology is only one-half of the equation. In order for technology to transition, a plan for transition must be developed and implemented. The overall concept for technology transition and the framework in which technology transition planning takes place are described in Air Force Materiel Command 80-50 series regulations. The focus is on orderly, phased transitions to give the user the technology option to satisfy part of his needs quickly, at an acceptable risk, in addition to having comprehensive quality technology available. Three aspects necessary for successful technology transition planning are organizational commitment (teamwork), technology availability for transition (timeliness), and operational requirements.

Teamwork across the Air Force Materiel Command product centers, particularly among laboratories, program offices, and development planning and engineering organizations, is the key to success in expeditiously transitioning technology into future weapon systems. Those specific processes for encouraging teamwork are established by the product division commander and managed as a corporate enterprise. Traditionally, the laboratory has had little control over or direct responsibility for technology application in systems. Accordingly, a measurement of Air Force laboratory success in readying technology for application is the completion of technology demonstrations planned in cooperation with one or more product divisions who agree to take the results and advocate the technology insertion into systems. These technology demonstrations have been described earlier as Advanced Technology Transition Demonstrations or ATTDs. The laboratories have the responsibility of developing a Technology Transition Plan (TTP) for all ATTDs. A TTP is an agreement between the laboratory, the product center technology transition focal point, and the technology recipient (i.e., System Program Office, Air Logistic Center (ALC), HQ Air Force, operational MAJCOM, support and training MAJCOM, and other equivalent DoD and governmental acquisition organizations and agencies). The TTP documents the specific tasks that must be successfully completed prior to acceptance by the technology recipient. Close cooperation is needed between SPOs and ALCs, the laboratories, and the appropriate MAJCOM customer in planning for technology transition.

Since most system development and acquisition are carried out through contracts with industry, the principal technology mechanism involves efforts by the laboratory to develop technology using one or more industrial firms who are familiar with and competent to employ the technology. These firms may then propose and execute designs embodying the technology through the System Program Office. If all parties are comfortable with the risks involved and are in agreement with the overall technical performance parameters, transition is smooth. In some instances, the special capabilities of the laboratories and the urgency of an operational requirement make it appropriate for the laboratory to develop products for direct delivery to a MAJCOM or other user. In these cases the laboratory makes an agreement directly with the technology user. This teamwork approach assures technology transition by obtaining user/developer commitment to a transition plan and by completing the technology demonstration on time for a scheduled transition opportunity.

Timeliness is the second factor necessary for successful technology transition. The primary mission of the Air Force laboratories is to develop and mature technology options for insertion into Air Force weapon and support systems. This is to provide demonstrated technology options, each with defined benefits and risks, which the MAJCOM, the product division, or other user can exploit in subsequent system application. The opportunities for transition are well understood by the developing organizations. Those transition opportunities are described in the technology transition plan. The laboratories are required by the Technology Executive Officer (TEO) to prepare a "baseline schedule" for each of their ATTDs. The "baseline schedule" must contain the objectives of ATTD, the major milestones both planned and accomplished, the transition opportunity (system and time frame), and the funding required to achieve the stated goals in the TTP. Each year, the TEO reviews each ATTD baseline to assess overall timeliness for technology demonstration to meet transition "windows of opportunity."

A technology transition plan developed by a team of acquisition experts with their eye on transition opportunities will ensure specific tasks, conditions, and criteria are sufficiently demonstrated to complete transition. The plan will include those options enabling a technology customer to apply some technology early as a partial solution to his needs. A TTP includes the criteria not only to reduce risk, but also to identify other factors related to further hardware and software development, producibility, and supportability to ensure a balanced technology transition package is available to the user.

The delivery to a user of a specific hardware, software, or information product is not the conclusion of the laboratories' responsibility to technology transition. The laboratories have the expertise and resources not only to meet a specific transition opportunity; they are capable of participating in other areas of the acquisition cycle more common to the program manager.

6.8   Science and Technology in the Acquisition Cycle

As described in Fig. 6.2, the laboratory management of RDT&E ends with the transition of technology into 6.3B or Demonstration/Validation. However, the laboratory can assist in many aspects prior to and after 6.3B. At the beginning of the acquisition cycle, when a need or requirement has not surfaced either within the MAJCOM community or the development planning office, the laboratory, in conjunction with the MAJCOM and the development planners, can perform technology tradeoff analysis that identifies potential military operational significance. With that identification, the laboratory community proposes new ATTDs that offer alternatives to current solutions. These ATTDs may be competing alternatives to existing technology demonstrations. However, as funding and available resources decline, competing alternatives may be a luxury.

After the definition of mission need, and before demonstration/validation, system concepts and analyses are performed to define relative risks of applying demonstrated technology and to lay an acquisition strategy appropriate to the mission need. The laboratories are advisors to the system program offices on technical matters and they are included on initial source selection decisions on basic concept definitions. Involving the laboratories early on in this phase can lead to a better understanding of the technologies and their integration risks as well as a reduction in overall design and manufacturing time.

Again, the identification and demonstration of advanced technology during this phase would be beneficial to technology transition and at the same time the cost, schedule, and performance associated with the technology can be assessed.

It is during the Demonstration/Validation (at Milestone I) phase where the proof of technology transition is displayed. At this time in the acquisition cycle, actual prototype hardware is produced. This hardware is usually one of a kind and sometimes handmade. In this phase of technology development, the critical design of a weapon system concept is assessed, and the confidence in the technology is established. The program manager is now responsible for demonstrating the integration of technologies that have come from the laboratories and from industry. The overall product goes through a set of rigorous testing to determine system performance parameters in a true operational environment. Based upon the results of this phase of testing, program risk is established. If the laboratories, systems engineers, and industry have done sufficient technology transition planning, the risk level will be low and technologies will be included in the final design. The laboratories play an important role in the assessment of system performance during the demonstration/validation phase in order to advise the program manager and the system engineers as to the performance tradeoffs and risks associated with major design changes. Additional feedback is provided the laboratories during the testing so that ongoing technology demonstrations can be modified to account for unforeseen complications or restructured to verify integration phenomena. In areas where technology is advancing so quickly, the laboratories, with their expert counterparts in industry and academia, can provide advice on the state-of-the-art at the time of design freeze so as to include only the latest technologies. The suggestion of a preplanned product improvement (P3I) can result if the system requires technologies that are rapidly advancing or if the threat is increasing.

Prior to the Engineering and Manufacturing Development (at Milestone II), the laboratories can provide the assessments needed to help systems engineers determine the overall risk of the concept going into the final engineering phase. It is also at this point that a recommendation for acquisition strategy is given to the program manager, that may include schedule changes, funding profiles, test criteria, and design changes.

At the point that a system has been selected to proceed into engineering and manufacturing development, several opportunities exist for the laboratory community to provide technology advice for insertion into a weapon system.

Although the focus changes from proof of concept to production, the laboratory works closely with the program manager to define areas in which further technology development or demonstration is required before full-scale production is begun. Demonstrations of maturing technology can be designed for which a preplanned product improvement is warranted, and longer range higher risk technology efforts such as those in basic research and exploratory development can be formulated and planned for future systems acquisition programs. The process generates those needs necessary for the laboratories to concentrate on within their program. The laboratories develop a Product Technology Plan in concert with their product division development planning organizations in four specific areas of Flight Vehicles, Space and Missiles, Human Systems, and C3I. These plans are different from the Technology Area Plans (TAPs) mentioned earlier in that the focus is on the system concept, the preplanned product improvement, or the future modification or block change scheduled. This planning document keeps the laboratories' technologies roadmapped toward system applications and ensures better transition opportunities.

Elements that must be considered in all technology programs before they reach the advanced technology transition demonstration stage are reliability, maintainability, supportability, testability, and manufacturing capability. Certainly before the technology can be certified for engineering and manufacturing development, the ability to test and manufacture is critical. The laboratories are now including in their test programs the requirement for design-to-test so that when the system goes to development test and evaluation, an assessment can be made of the performance characteristics. As is the case with integrated avionics, where the functions are constantly changing, it becomes a test engineer's nightmare to know how to test the technology or even how to know if the data are valid. Additionally, the ability to manufacture the technology is being included in the development and demonstration phases of the technology within the laboratory. These process oriented technologies are extremely important in technology development where there are changes not only to structures within weapon systems, but also to the materials load-carrying parameters and radar cross sections.

6.9   Science & Technology Program Objectives

The objective of any defense S&T program is the continuing discovery, exploitation, demonstration, and rapid transition of technology into quality weapon systems to meet operational needs. In order to do this, our defense laboratories, their product division engineers, and industry must work together to plan, demonstrate, and deliver timely technology that is matched by no one. What has been described here for the case of Air Force is a disciplined process for structuring technology transition and transfer. It is fundamental to the United States remaining first as an aerospace nation. More than at any other time in our history, the defense S&T programs must operate effectively at reduced budgets and at the same time must continue to create the necessary technological edge for our present and future weapon systems.

The Stevenson-Wydler Technology Innovation Act of 1980 (PL 96-480), as amended by the Federal Technology Transfer Act of 1986 (PL 99-502), and Presidential Executive Order 12591 "Facilitating Access to Science and Technology," April 10, 1987, encourage the transfer of technology from Federal laboratories to State and local governments and to private sector. The Department of Defense fully supports and promotes the use of knowledge from the research laboratories to develop new products with potential application in the private as well as the public sector.

References

[1] White House, "National Security Strategy of the United States," Washington, DC, Government Printing Office, 1991.

[2] Chairman, Joint Chiefs of Staff, The Pentagon, "National Military Strategy," Washington, DC, Government Printing Office, August 1992.

[3] Director, Defense Research and Engineering, Department of Defense, "Defense Science and Technology Strategy," Washington, DC, July 1992.

[4] Director, Defense Research and Engineering, Department of Defense, "DoD Key Technology Plan," Washington, DC, July 1992.

[5] Przemieniecki, J.S., Ed., "Critical Technologies for National Defense," Education Series, American Institute of Aeronautics and Astronautics, Washington, DC, 1991.

[6] Rankine, R.R., Maj. Gen.,"Total Quality Treatment for Science and Technology," Aerospace America, American Institute of Aeronautics and Astronautics, Washington, DC, pp. 36-40, May 1992.

Chapter 7

FINANCIAL MANAGEMENT

7.1 Introduction

Financial management is one of the least understood areas in the acquisition arena, and is also one of the most publicized topics because of the high stakes in the annual authorization and appropriation process. There are four activities, or phases, associated with financial management: budgeting, enactment, execution, and financial reporting. This chapter will examine the first three phases of the process, while the next chapter will address the reporting aspects of financial management.

7.2 DoD Resource Management: The Biennial Planning, Programming, and Budgeting System (BPPBS)

The Biennial Planning, Programming, and Budgeting System (BPPBS) is very complex, impacting everything the DoD and military services do. Planning is very important–it lays the baseline. Programming is very important–it shapes the system. But we must remember that the end product is the budget. We will now take a look at the history of the BPPBS, and then examine each of these activities in greater detail.

The BPPBS started as the Planning, Programming, and Budgeting System (PPBS), which was introduced in the DoD in the early 1960s by former Secretary of Defense McNamara. At that time, the Department of Defense (DoD) received over 50 percent of the U.S. budget. Today, priorities have changed so that the DoD receives less than 20 percent of the U.S. budget, and the amount continues to dwindle annually, indicating a significant shift in national priorities.

McNamara created the PPBS to formalize the decision process for weighing the costs associated with major weapons systems. The PPBS's principal purpose was to rationalize planning and budgeting functions and to add a new function, programming, to provide a bridge between planning and budgeting. The PPBS was used to provide a Five-Year Defense Program

(FYDP), a automated data base of all Defense programs, approved by the Secretary of Defense. The FYDP was based on submitting budgets annually, with the first year being the budget year. In 1985 the PPBS became the BPPBS when Congress tasked the President to submit 2 year budgets for the DoD to provide more continuity in program funding. This has had very little effect since Congress still only approves annual budgets. This also changed the FYDP, which was then known as the Six Year Defense Program, or SYDP, where the first 2 years are the budget years. Sometime in 1990 the SYDP was renamed back to the FYDP, only this time FYDP stands for Future Years Defense Program. We'll talk more about the FYDP later in this chapter, and now we will look at the BPPBS in greater detail.

BPPBS PHASES

The BPPBS is an iterative process that involves three separate management phases: planning, programming, and budgeting (see Fig. 7.1).

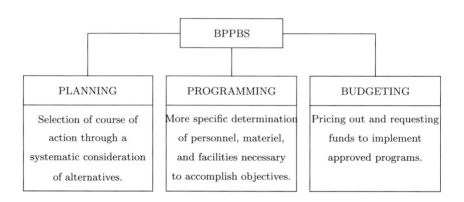

Fig. 7.1: Management phases of the BPPBS system.

Planning: Planning involves the systematic consideration of alternative actions to accomplish DoD objectives out as far as 15 years beyond the current FYDP. The planning function provides the framework for subsequent DoD programming and budgeting actions because it is primarily goal or output oriented. The objective in this phase is to identify strategies and capabilities which DoD must develop to support national security objectives. The primary constraint is that what is recommended must be achievable.

Two of the documents discussed in the BPPBS process, the National Military Strategy Document (NMSD) and the Defense Planning Guidance (DPG), are the outputs of the planning phase. These documents were also discussed in Chapter 1.

Programming: Programming involves reviewing and ranking the capabilities which DoD must develop, and subsequently attempting to program resources to match these strategies. The programming process is a systematic review and consideration of the currently approved programs as expressed in the FYDP, along with an evaluation of new program alternatives, and extends for up to 9 years from recommendation submission.

The competition for funds dictates that some strategies may not be feasible when resources are considered. As a result, the programming process may involve some of the same activities as the planning process, since priorities may have to be changed or strategies modified. Programming provides the bridge that ties the planning function (the overall mission) with resources (dollars, personnel, etc.)

Documents associated with programming are the Program Objective Memorandum (POM), Issue Papers (IP), and the Program Decision Memorandum (PDM).

Budgeting: Budgeting is the process by which program decisions are translated into appropriations requests. It is the process by which we identify funding requirements for the President's Budget and Congressional action.

Like programming, budgeting is a function that sometimes involves elements of the preceding functions. As budget decisions are made, it may be necessary to reconsider strategies previously developed or to reschedule previously approved programs. Since many months pass in the completion of one BPPBS cycle, it is inevitable that plans must be altered and programs adjusted to reflect prudent responses to the changing defense priorities.

The primary outputs, or documents, of the budgeting process include the Budget Estimate Submission (BES), Program Budget Decisions (PBDs), and the President's Budget.

BPPBS PROCESS

Now that we have examined the three phases of the BPPBS process, we will look at how they fit together and the documentation that goes along with the process. The BPPBS is how each military service manages inputs and outputs of the FYDP. The BPPBS cycle is the sequence of events used to accomplish the planning, programming, and budgeting functions. The

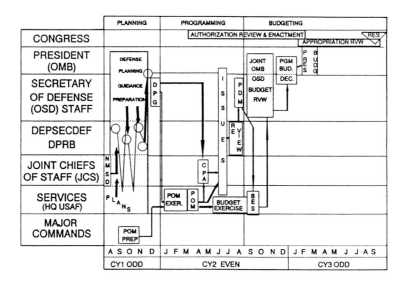

Fig. 7.2: Biennial Planning, Programming, and Budgeting System cycle.

cycle is based on a calendar which provides major milestones beginning as much as 3 years prior to the start of the budget fiscal year. Because this cycle consumes more than 2 years, an odd-year/even-year perspective is useful in developing some understanding of the process. Fig. 7.2 shows the overall BPPBS cycle.

ODD-NUMBERED YEAR

We will start this discussion at the left-hand side of Fig. 7.2, and put it into a time frame that we can work with. The beginning point that we will use is October 1991, which is the beginning of Fiscal Year (FY) 92. Fig. 7.2 is based on calendar years, and not fiscal years, so we will have to do some translating. We know that the President delivered the budget request for FY 92/93 to Congress in January 1991, and we will use that as a base to work from also. We can see that the National Military Strategy Document (NMSD), which actually occurs a little earlier than indicated in this figure and kicks off the BPPBS cycle, has already been released. We can also see that the Draft Defense Planning Guidance has been issued, and we are getting ready to release the Defense Planning Guidance (DPG) to the various DoD components. One question you may be asking yourself is what years this cycle will cover. We know that the President delivered his FY 92/93 President's Budget (PB)budget request in January 1991, and that he is getting ready to deliver his FY 93 Amended President's Budget (APB) request in January of 1992, both of which fall out of the current planning phase. With this information, it becomes obvious that the BPPBS cycle starting in 1991 covers FY 94 and FY 95 for budgeting, and FY 96 to FY 99 for the rest of the Future Years Defense Program.

Step 1: National Military Strategy Document (NMSD): The NMSD is developed in the Office of the Joint Chiefs of Staff (JCS), and provides the Secretary of Defense with a statement of recommended military objectives derived from national objectives, and the recommended military strategy required to attain them. Included is a summary of JCS planning force levels which could execute, with reasonable assurance, the recommended strategies. Also, included are JCS views on the attainability of the force levels. The NMSD is submitted to the Secretary of Defense (SECDEF or OSD) in early September and provides the foundation for JCS recommendations on force planning guidance and changes to the Defense Planning Guidance.

Step 2: Defense Planning Guidance (DPG): The DPG, prepared by the Defense Planning and Resources Board (DPRB), and issued by the Secretary of Defense in December, provides fundamental policy, strategy, issues, and rationale underlying the total defense program, and high level Program Objective Memorandum (POM) guidance. The JCS, Services, and Defense Agencies receive a draft DPG in early October. It is intended that a dialogue occur between the Services, the JCS, and the OSD staff before the issuance of the final version.

EVEN-NUMBERED YEAR

As we move to the right in Fig. 7.2, we go into the even calendar year, in our case 1992, and move into the programming and budgeting phases of the BPPBS cycle. You might recall that we said in the last section that the President will deliver his amended budget request for FY 93 to Congress in January 1992. Since we develop budgets biennially, as directed by Congress, but Congress only acts on annual budgets, what do we do in the off years? Basically, in the off years we look at changes that have occurred from the original cycle, identify disconnects, and then work them into a revised budget in much the same way as we did the original, two-year, budget.

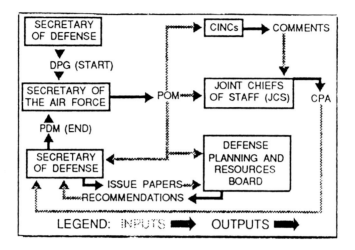

Fig. 7.3: POM review cycle.

Step 3: Program Objective Memorandum (POM): In April or May each military department and Defense agency submits a POM to the Secretary of Defense. The POM presents a priority-ranked program and includes baseline force levels, support and activity levels, and deployments within the stated constraints and fiscal levels of the Defense Planning Guidance (DPG). The POMs include an analysis of each proposed change or new program. They express force, manpower, and cost implications for the two budget years plus four additional years. Several of the subsequent steps involve POM review cycle, as shown in Fig. 7.3.

Step 4: Chairman's Program Assessment (CPA): The CPA provides the views of the JCS on the adequacy and capabilities of the total forces contained in the Service POMs to execute the national military strategy, and the risks inherent in those force capabilities. The CPA includes:

1. An assessment of the capabilities and associated risks represented by the composite POM forces.

2. The JCS view on the balance of the recommended Service force and support levels.

3. JCS recommendations on how to achieve improved defense capabilities within the alternate funding levels directed by the Secretary of Defense (SECDEF or OSD).

4. A mobility force analysis.

Step 5: Issue Paper (IP): During the June-July period the Secretary of Defense conducts a detailed review of the Service programs and makes decisions based on the POMs and Issue Papers. The IPs define specific issues for review by comparing the proposed program with the objective and requirements established in the DPG.* The papers, which are consolidated into Issue Books for transmittal and review, present alternative resolutions for each issue and evaluate the merits of the alternatives, in terms of the DPG fiscal constraints and their ability to implement the missions set forth in the DPG. Issue Papers are prepared in the following categories:

- Nuclear forces.

- Conventional forces.

- Modernization.

- Intelligence.

- Manpower.

- Logistics and readiness.

Step 6: Program Decision Memorandum (PDM): In late July or early August a PDM is issued by the Secretary of Defense (SECDEF or OSD) for each military department and defense agency. These PDMs summarize the initial program decisions of the current cycle based upon a review of the POMs, CPA, and IPs. The PDMs are strongly influenced by the deliberations, recommendations, and decisions of the Defense Planning and Resources Board (DPRB). They constitute budget guidance to the recipient organizations as those organizations prepare to submit their budgets. The Program Decision Memorandum limits the items that a Service can submit budgets for by paring down the POM input to the mixture of programs that the DoD can afford to fund based on priority.

Step 7: Budget Estimate Submissions (BES): This step puts us into the final phase of the BPPBS, budgeting. Moving further to the right in Fig. 7.2, we see that we are rapidly approaching the final steps getting our requests into the FY 94 Presidential Budget. In mid-September, each military department and defense agency submits its annual budget estimate to

---

*Examples of current issues are the relationships of Reserve to Active Duty Forces, and conventional to nuclear weapon capabilities.

the Secretary of Defense. The estimates update the approved programs reflected in the PDMs. Specific detailed instructions for budget submissions are prescribed by OSD, in accordance with Office of Management and Budget (OMB) directives. Upon receipt of the budget estimates, the Secretary of Defense directs a review by the OSD staff working with representatives of the Office of Management and Budget.

The current cycle considers both the FY 94 and 95 budgets, but only the FY 94 budget was mentioned in the last paragraph. Why? Although DoD is on two year cycle budgets, Congress deals with annual budgets. In September 1993, the various DoD components will submit an amended BES followed by the President's Budget for FY 95.

Step 8: Program Budget Decisions: In late October the first Program Budget Decisions (PBDs) are issued. Departments, JCS, and agencies provide comments on an "as received" basis in order that the last budget decision may be issued by mid-December. Specific budget issues that arise may result in Major Budget Issue (MBI) meetings with the Secretary of Defense (SECDEF or OSD).

Step 9: DoD Budget Submission: Late in December the completed DoD budget is sent to the Office of Management and Budget for review and presidential approval. When approved it becomes a part of the President's Budget submitted to the Congress.

ODD-NUMBERED YEAR

We are now at the far right side of Fig. 7.2 and, based on our example, into January 1993, when the President will submit his budget request for FY 94. This constitutes the final step of the BPPBS process.

Step 10: The President's Budget: In January of each year the President submits a Unified Federal Budget to the Congress. While the President submits an annual budget for all other executive branch agencies, the DoD submission is biennial (January of the odd-numbered years). The even-year DoD portion of the President's Budget requests necessary budget adjustments to the second year of the DoD submission which was made the previous year.

DEFENSE PLANNING RESOURCES BOARD

One of the organizations listed in Fig. 7.2 was the Defense Planning Resources Board; something we will look at in greater detail. The Defense

Planning and Resources Board (DPRB) helps the Secretary of Defense manage the entire BPPBS. The DPRB reviews the proposed planning guidance, manages the program and budget review process, advises on PPB major issues and proposed decisions, and assures the alignment of major system acquisitions with available resources. DPRB members represent the services, the JCS, and OSD staff functions. The Deputy Secretary of Defense is the Board Chairman.

FUTURE YEARS DEFENSE PROGRAM (FYDP)

The FYDP is the official, formal written record of all financial management decisions that have been made in the DoD since 1962. Its purpose is clearly defined in Department of Defense (DoD) Directive 7000.1, which states that the FYDP will contain DoD approved plans. The FYDP, shown in Fig. 7.4, is identified as "the nucleus of the Department of Defense resource management systems," and systems developed for resource management "will be consistent with it." To maintain this consistency, the FYDP is updated five times during the PPBS cycle through a series of exercises. The "A" series exercises provide an update based on the POM. The B-series reflects an update based on the BES, and the C-series update reflects the President's Budget.

You may recall that we said earlier in this chapter that the FYDP is a computerized data base, and is capable of retaining massive amounts of data. This section explains the nature of the data captured in the FYDP including the types of data maintained (historical and projected), and how the data are classified within the FYDP structure.

Data Classification: The FYDP identifies each piece of data using three separate classification schemes. The data are classified by input (appropriation), by processor (Military Department or DoD Agency), and by output (Major Force Program). The result is a matrix, so that the data can be accumulated along whichever lines are most useful to the decision maker. We will now examine how the FYDP is organized to account for these resources.

Outputs: The outputs of the FYDP are classified in terms of Major Force Programs. FYDP designers recognized that any attempt to describe DoD outputs with any degree of precision would result in a detailed and cumbersome structure which would be difficult to comprehend. Additionally, if the descriptions were too specific, it would be difficult to develop a presentation which unified the military departments and DoD agencies. Therefore, descriptions had to be very broad.

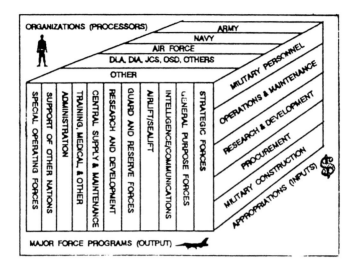

Fig. 7.4: Future Years Defense Program structure.

Initially nine major programs were developed to describe outputs. Various modifications have led to the current structure of eleven major force programs:

1. Strategic Forces.

2. General Purpose Forces.

3. Intelligence and Communications.

4. Airlift/Sealift.

5. Guard and Reserve Forces.

6. Research and Development.

7. Central Supply and Maintenance.

8. Training, Medical, and Other General Personnel Activities.

9. Administration and Associated Activities.

10. Support of Other Nations.

11. Special Operations Forces.

These programs are general in nature and describe activities in which all elements of the Department of Defense are involved, yet they are specific enough that the terms are meaningful to FYDP users. For example, the term "Strategic Forces" would invoke images of an intercontinental weapons delivery capability. Within this major program, you expect to find specific capabilities such as the Air Force's long-range bombers and intercontinental missiles, and the Navy's sea-based missile systems. By defining outputs in this manner, DoD decision makers can review alternative means of achieving a capability with the ability to easily analyze across service lines.

As previously identified, the FYDP has to be useful to planners. Since planners deal primarily in terms of capabilities, it is necessary to include information on approved force levels. While the number of organizations and types of weapons systems are identified, the FYDP does not identify exact numbers of weapons systems. Quantities are, however, identified for new acquisitions. Thus, the FYDP may identify eight squadrons of B-2s programmed for FY 9X, but will not identify the exact number of B-2s to be operational. It may, however, identify that seven B-2s are programmed to be acquired in FY 9X.

Program Elements: Program Elements (PEs) are subdivisions of Major Force Programs that identify a specific capability. While the Major Force Programs are useful general descriptors of DoD output, decision makers need specific information on the weapons systems that contribute to the general capability. This specific information is identified with a program element, a subdivision (or an element) of a Major Force Program. Thus, the program element is the basic building block for describing the output of the DoD. Every mission of the DoD must be incorporated into one of the eleven Major Force Programs and in turn into one of the many Program Elements.

A program element is defined as "a combination of personnel, equipment, and facilities...[which]...constitutes a military capability or support activity." Thus, within the Strategic Forces Major Force Program, you would find a program element identifying the B-1 capability and another program element identifying the PEACEKEEPER missile capability. (NOTE: There is one program element for the entire capability and not a separate program element for each B-1 or PEACEKEEPER squadron.) It is obvious that these program elements provide a very specific description of the outputs of the DoD.

The decision maker need not rely solely on the Major Force Program grouping if a rearrangement of program elements is more appropriate for the alternatives being studied. Additionally, tradeoffs can be analyzed comparing various "mixes" of forces within any Major Force Program.

Inputs: DoD Directive 7000.1 requires that the Future Years Defense Program contain a language such that "...budgeting...will be consistent with it." Budgeting is a process to identify and obtain spending authority from Congress; thus, it focuses on what will be purchased with the spending authority (i.e., the inputs to the DoD production process). Since the FYDP needed to accommodate budgeting, its structure had to include input information.

Congress makes "inputs" into the Department of Defense budget process by deciding how much authorization/appropriation authority the department should have. Remember, the DoD only makes budget requests. Congress approves the budget. Once DoD receives its annual appropriation authority from Congress, DoD releases appropriation authority to the services for the current fiscal year. For instance, MILPERSONNEL (military personnel) appropriations are spread across all of the 11 major force programs to pay military salaries and personnel expenses. DoD converts an input from Congress, called an appropriation authority, into an output, or one of the major force programs.

Processors: The conversion from inputs to outputs takes place in the organization responsible for the conversion. Since this organization is closest to the production process, its managers are responsible for identifying what inputs are required to produce the outputs. Additionally, they are responsible for the performance of their organization in producing outputs effectively and efficiently. The Military Department or DoD Agency identifies which DoD organizational element carries out the conversion process. With the FYDP data base maintained at DoD level, the organizational identification goes no lower than department or agency level. At lower levels within each agency, responsibility centers and accounting centers are specifically identified.

Historical Data: The FYDP serves as a data base for DoD financial information retaining all data on DoD programs since the establishment of the FYDP in 1962. The data can be used to analyze trends and to assist with projecting future requirements. One major problem in developing such analyses is related to the fact that the FYDP structure has not been static, nor should it be. As it is true of any database, the FYDP is adjusted

to meet the needs of users for data in specific formats and combinations.*
The FYDP began with nine Major Force Programs (MFPs) and expanded
to eleven. As program definitions have changed, the alignments of program
elements have changed. Additionally, new program elements are constantly
being added as new capabilities are designed. Each of these factors hampers
the ability to make valid comparisons using different years of historical data.

Projected Data: Since the Future Years Defense Program is a data base
of all DoD financial decisions, we must have decisions that are made about
the future as well as those that have been made in the past. Time intervals
for the projected data varies by category.

Projected Costs and Personnel: For development of resource projections,
the FYDP is truly a 6-year program because it retains cost and personnel
projections for 6 years into the future. This data is termed "priced" data
since it identifies quantities of input (i.e., prices) necessary to produce output.

Projected Forces: Since planners necessarily work on a relatively long
horizon, the FYDP retains force projections for 9 years into the future.
This longer time frame is necessary to accommodate the longer lead times
often required for developing and acquiring new systems.

7.3 Enactment, Authorization, and Appropriations

ENACTMENT

Once we have completed the first phase, the budgeting development process,
through submittal of the President's Budget, we are ready to enter into the
second phase of the financial management process, where Congress has to
provide legislation to give us authority to accomplish programs and funding
to run the programs. This is known as the budget enactment process.

Enactment basically consists of two things, authorizations and appropriations, both of which we need to be able to execute a program. Once
Congress receives the President's Budget, they start to work to build the
authorization and appropriation bills that will provide us the authority to
execute our programs.

The first thing that Congress is supposed to do is to draft a Concurrent
Budget Resolution that sets federal spending and revenue targets, or limits,
for the upcoming fiscal year. The Concurrent Budget Resolution for those

---

*The most recent addition of MFP 11, "Special Forces Operations", is a good example
of such changes.

reasons is often called the "Congressional Budget" to contrast it with the budget proposed by the President. This resolution does not have the power of law, however, and is often ignored by the spending committees requiring eventual reconciliation of the entire process.

AUTHORIZATION

An authorization is the Congressional authority to carry out a particular program within specified restrictions. Basically, Congress gives authorization to purchase certain equipment, products, services, or weapon systems, assuming funding is appropriated. There are two authorization bills that have to be approved by the Senate Armed Services Committee (SASC) and the House Armed Services Committee (HASC): the Defense Authorization Bill and the Military Construction Authorization Bill. These bills should reflect the overall spending limits set in the Concurrent Budget Resolution, but often don't. In addition, before they are completed , the appropriation process is well under way.

APPROPRIATIONS

Appropriations are permission to obligate the Treasury to pay money for goods or services. The Appropriations Bill passed by Congress authorizes the obligation and subsequent outlay of certain dollar amounts for specific purposes. The Senate Appropriations Committee (SAC) and House Appropriation Committee (HAC) also have two bills that they must draft, the Defense Appropriation Bill and the Military Construction Appropriation Bill. The appropriation bills should reflect dollar amounts and quantities less than, or equal to, those shown in the authorization bill. Occasionally appropriations are received without authorization, and the issue will have to be settled prior to obligation.

In 1950, the number of appropriations was reduced from 2,000 to 375, with each appropriation covering a broader range of items to be purchased. The Department of Defense annually receives approximately 85 appropriations which fall into five general areas: Military Personnel; Operation and Maintenance; Procurement; Military Construction; and Research, Development, Test, and Evaluation.

Each of these general areas represents several appropriations. For example, in the area of operation and maintenance, the Air Force alone receives three appropriations, one for day-to-day operations of the active duty establishment, one for operation of the Air National Guard, and one for operation of the Air Force Reserve. Altogether, there are 12 to 15 annual appropriations for operations and maintenance of DoD activities. Much as the Major Force Programs broadly describe outputs, the appropriation categories provide a very general description of inputs. These general descriptions, however, provide categories useful for Congressional review and decisions.

Continuing Resolution Authority: Ideally, the Authorization and Appropriation bills are passed by Congress, and signed by the President, prior to the beginning of a new fiscal year on October 1. In reality, what normally happens is that Congress has to draft a Continuing Resolution Authority (CRA) by October 1, that allows obligation, within specified parameters, until Congress and the President can agree on final authorization and appropriation bills.

A Continuing Resolution Authority allows us to conduct business prior to signing the bills, but with certain restrictions. For example, new starts and new multi-year procurement cannot be initiated. The second restriction is that funding quantity increases above previous years levels are not authorized. The final restriction is that programs are not allowed to obligate at levels higher than 80% of the lowest Congressional Committee "marks" or the previous year's funding levels, whichever is lowest. The term "mark" refers to the proposed program budget in all four committees, the SASC, HASC, SAC, and HAC.

7.4 Execution

Once the authorization and appropriation bills are passed by Congress and signed by the President, the Services must decide how to manage this money in the form of appropriations. This is the third phase of financial management – Execution. Program managers must decide, within the constraints included in the laws, on the most efficient and effective use of these funds.

Funds Status: There are several stages of accountability of appropriated funds in the execution phase: commitment, obligation, and expenditure.

Commitment: Funds are committed when someone in program control agrees to ear-mark them for a certain purpose and Accounting and Finance has certified fund availability. This fund commitment is usually done in response to a new initiative on a program, and is normally done against the management reserve that a Program Manager has to work with. Different valid needs usually compete for limited dollars, but getting a commitment of funds is the first step in getting new initiatives going. An example of a commitment is the use of the Air Force Form 9 to administratively reserve funds.

Obligation: Funds are considered to be obligated after a formal contract has been entered into. It is the process of getting those previously committed funds on a legal contract to obtain work for the program office. As we discuss later, funds expire after a certain time duration. If the funds are not obligated prior to expiration, they are lost forever, except to cover claims that may arise over previous obligations.

Expenditure or Outlay: This is the act of putting the check in the mail, or indicates that we have forwarded the obligated money to the contractor who has completed the work.

APPROPRIATION CATEGORIES

As we discussed in the FYDP section, there are different types of appropriations, each with varying lengths on the time for obligation. There are two things we have to be concerned with in the execution phase, fiscal integrity and appropriation integrity. For example, we cannot spend money on aircraft spare parts with money earmarked for military construction, and we cannot obligate funds for new efforts after they have expired.

It is important to understand each appropriation and the time available for obligation. As we look at the time we have available for obligation, we should remember that these time-spans represent maximum times. Normally, if program funds are not obligated before the expiration date, the funds are subject to reprogramming for higher priority requirements.

3010/3020/3080 Appropriations: 3010/3020/3080 appropriations are used to procure systems and have an Obligation Availability Period (OAP) of three years. 3010 appropriations are for aircraft, 3020 for missiles, and 3080 is for other procurement (such as vehicles, conventional munitions, etc.).

3300 Appropriation: 3300 appropriation is used for military construction and has an OAP of five years.

3400 Appropriation: 3400 appropriation is for operations and maintenance, is good for one year, and includes funds for operating expenses such as supplies, TDY, and civilian personnel salaries.

3500 Appropriation: 3500 appropriation is used to pay for military personnel and has an OAP of one year.

3600 Appropriation: 3600 appropriation is used for research, development, test and evaluation of new systems, and has an OAP of two years.

EXPIRED APPROPRIATIONS

Prior to 5 December 1990, those funds that were not obligated were placed in a surplus authority account for two years, maintaining both their appropriation and fiscal identities. After these two years, the unobligated funds moved into a merged surplus authority, where they maintained the appropriation identity, but lost any fiscal identity, and could be restored to an "M" account to pay for bona fide adjustments to existing contracts. A bona fide adjustment is one that does not change the scope of the existing contract or increase quantities ordered.

Funds that had been obligated, but not expended, followed a different track. After the time period allowed, funds that were obligated, but not expended, were placed in an expired status and tracked by the individual services, where they retained their appropriation and fiscal integrity for two years, and then were maintained by the individual services in "M" accounts, where they maintained their appropriation identity and lost their fiscal identity.

When the law covering this expired funds process was first drafted in 1956, it was envisioned that the balances in these accounts would remain relatively low. Unfortunately, this was not the case, with the combined services merged surplus authority rising from $5.2 billion in 1980 to $27 billion in 1990, and the combined services "M" accounts rising from $2.7 billion in 1980 to $18.8 billion in 1990. The use of these accounts came under Congressional scrutiny in 1989, when the Air Force informed Congress that it planned on using almost $1 billion from the accounts to correct deficiencies on the B-1B.

These events led to new legislation in the FY 91 Authorization Act, and the gradual phase-out of "M" accounts by September 30, 1993 (PL 101-510). Money that was in the surplus authority for appropriations that expired in FY 89 and FY 90 was transferred to the agencies' expired accounts. Each agency must now maintain their own balances of unobligated and obligated expired budget authority that will retain both appropriation and fiscal identities. The expired period of tracking was increased from two years to five years, and at the end of 5 years, all funds that are still unobligated or unexpended will be canceled.

This new legislation also eliminated the merged surplus authority that the Treasury kept indefinitely. All "M" accounts will be canceled as of September 30, 1993, and funds transferred to the "M" accounts prior to that date will be subjected to the same 5 year rule (i.e., if the money that is in the "M" account is more than five years old, it will be canceled).

FINANCIAL ACTIVITY OVERLAP

It is clear from the previous discussion of the various phases of the financial management process in the Department of Defense that several cycles of the process are simultaneously in progress. In fact, there are five biennial cycles in progress at any one time. Also, the following five activities are involved in the process for at least one cycle at any time:

- Planning

- Programming

- Budgeting

- Enactment/Authorization

- Execution

The cycle overlap of this process is shown in Fig. 7.5.

7.5 Future Changes

This chapter discussed the processes by which our nation decides how much of and for which purposes our national resources will be allocated to DoD requirements. BPPBS has been and will remain, in the foreseeable future, the system through which the DoD identifies these needs. These needs are incorporated into the President's budget proposal, through the Unified Federal Budget, to the Congress. However, the chapter has not addressed a recent change, *internal to the Department of Defense*, which will modify the process by which most *support* units will acquire their funding. This significant change is the Department of Defense Business Operations Fund, more commonly known as DBOF.

Although the first stages of DBOF were implemented in FY92 (incorporation of the Stock and Industrial Funds), its final configuration is still to be determined. Consequently, no description of its processes was included in this chapter. One general statement is appropriate. For those business activities included in DBOF, the BPPBS will be used to acquire direct congressional appropriations for those organizational capabilities that are

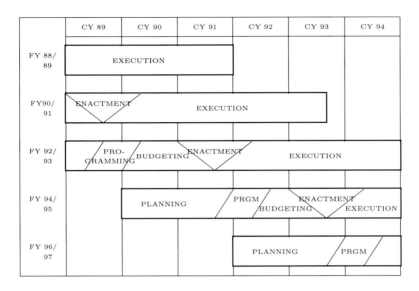

Fig. 7.5: Financial Activity Overlap (Planning, Programming, Budgeting, Enactment, and Execution).

required exclusively for readiness and similar war-fighting activities. For *support-type* organizations, this will constitute a relatively minor part of their total budget. Most of these funds required by such organizations will be earned by providing products and/or services to *mission* organizations who will continue to use the BPPBS as described.

References

[1] The Biennial Planning, Programming, Budgeting System (BPPBS)-A Primer, 7th Edition, published by HQ USAF/PE, January 1993.

[2] DoD Instruction 7045.7, Implementation of the Biennial Planning, Programming, and Budgeting System (BPPBS).

Chapter 8

PROGRAM CONTROL

8.1 Introduction

This chapter focuses on the important area of financial management – Program Control. Program control helps in the financial management process by providing the necessary information with which to make important programmatic decisions. Program Control assists other functional directorates in a typical System Program Office involved in costing and scheduling. This chapter introduces the common types of work found in Program Control, some of the tools used to track costs and schedule, the characteristics of cost estimating, and some of the reasons for weapon system cost growth.

8.2 Program Control and Cost Estimating

THE PROGRAM CONTROL ORGANIZATION

Typically, the individuals in the Program Control office are the "eyes and ears" of the Program Manager. They may not understand all the tasks like an engineer or a configuration management person, but they know the financial aspects of the program.

The Program Manager is the single point of responsibility in acquiring and deploying weapon systems to satisfy operational needs. The Program Control Directorate attempts to facilitate the execution of systems within schedule and acceptable cost constraints while meeting performance and logistic supportability requirements. Program Control personnel frequently develop key contacts in and out of the Government to carry out their responsibilities. They essentially filter information to the Program Manager so that issues and challenges are elevated and answered quickly and thoroughly. Sometimes informal contacts may be the best avenues for controlling costs, schedule, performance, or supportability.

Functionally, Program Control is typically divided into three unique divisions, depending on the type of work involved: The Program Evaluation which deals with the contractor, while monitoring the costs of the program; the Financial Management which focuses on identifying and monitoring funds to keep the program moving; and the Plans and Integration which deals with non-financial aspects of the System Program Office (SPO) like planning, scheduling, and forecasting. The following is a summary analysis of the special considerations of each Program Control Division.

Program Evaluation: Program Evaluation is really program analysis of costs in support of the financial manager. By interfacing with the financial managers, Program Evaluation identifies the cost requirements and the cost performance of the contractor. Some activities normally found in Program Evaluation are cost/schedule control systems criteria (C/SCSC) and cost analysis. C/SCSC deals with contractor performance measurement using Cost Performance Reports (CPR), Contract Funds Status Reports (CFSR), and Cost/Schedule Status Reports (C/SSR) used to track and analyze the contractor's cost performance. The emphasis in cost analysis, from a Program Control perspective, is in comparing the contractor's Estimate At Completion (EAC) to the SPO's budget. Some of the end products of Program Evaluation are cost estimates in source selections, periodic review of the SPO estimate for sufficiency, Program Objective Memorandum (POM) or Budget Estimate Submission (BES) inputs to financial management, budget support with financial managers, future billings and budget forecasts, and support for impact statements and "what-if" exercises in budget and briefing cycles.

Financial Management: Financial Management, on the other hand, deals with the management of program funds, the process of budget requests, and the review and status of prior year obligations. Some feel that this is the heart of any program because if there were no funds to run your program, you would not have a program to run. Typical Financial Management functions are budgeting, financial analysis, fund administration, fiscal integrity and accountability, appropriation integrity, and support of the Program Manager. Fiscal integrity, for example, means using FY90 dollars for the approved FY90 program, FY91 dollars for the FY91 program and so on. Appropriation integrity means, for example, that 3010 money is used to procure aircraft or that 3020 money is used to procure missiles.

Financial management end products include the POM and BES, fund status, funding documents, and the program baseline. The POM identifies total program requirements for the next 6 years, and includes rationale in support of the planned changes from the approved Future Years Defense Program (FYDP) baseline. The POM is based on strategic concepts and guidance stated in the Defense Planning Guidance and includes an assessment of the risk associated with current and proposed forces. The BES is a recosting of the POM as modified by the Program Decision Memorandum (PDM). It should be noted that new requirements cannot be introduced in the BES, only in the POM. Another end product of financial management is the status of funds. Weighing the contractors EAC to the SPO's budget is a continuous activity. At any time, a financial analyst is able to provide the status of funds with regard to expenditures and obligations.

Plans and Integration: Plans and Integration functions include scheduling, documentation review, program analysis, computer support, and reports control. A key element in scheduling is the integrated master schedule. An integrated master schedule is a detailed program schedule that portrays all of the major elements of a program and all related development efforts so that the interrelationships are easily seen. Since it is the integrated schedule, it is updated regularly and is recognized as the only authorized source for publication of schedule information outside the program office.

The other functions of Plans and Integration deal with supporting the Program Manager in non-financial aspects. For instance, documentation reviews are held to analyze all descriptive information concerning activities between Federal agencies and/or the contractor. These documents usually delineate policy or elevate issues so that options can be addressed. In a system of plans, the Program Manager's, and the team's, strategy to achieve the objective is integrated in the Program Management Plan and detailed in plans such as the Acquisition Plan, Test and Evaluation Master Plan, Integrated Logistics Support Plan, and lower level functional plans. End products associated with Plans and Integration include the master integrated schedule, internal reports, and ensuring that the Program Management Directive is valid and current.

Plans and Integration also deals with outside agencies like the Inspector General (IG), the General Accounting Office (GAO), and any other external agencies that may be seeking information. Plans and Integration acts as the focal point for the outside agencies, and will then coordinate with the appropriate individuals in the SPO to provide the requested information.

PROGRAM CONTROL AND THE BUDGET PROCESS

In previous chapters we talked already about major activities in the BPPBS process. The Program Control office is responsible for keeping track of the budget for the specific program. Next we will address the four areas of the budget process.

Program Control's involvement is most extensive in budget formulation since this phase is where the SPO budget is formulated. Ideally, the budget is based on the required costs necessary to satisfy program requirements. Cost estimating is the important activity in the budgetary process because this is where costs are developed and identified. Usually SPO budgets are based on validated cost estimates or cost estimates that have been reviewed and approved based on consistency, completeness, reasonableness, and documentation. Once the budget is formulated, the Program Manager agrees to deliver a weapons system for a certain cost which is documented in the program baseline.

The second phase, enactment, is getting authorization and appropriated funds from Congress designated for a specific program. The best way to justify program requirements is through good documentation. Since Program Control is directly involved in this process, good documentation can highlight the importance of the program in defending the program requirements before Congress.

The execution phase involves putting appropriated funds to work through commitment, obligation and expenditure. Program Control must ensure that fiscal and appropriation integrity are maintained in this process.

The fourth budgetary area is analysis and reporting. The analysis and reporting phase involves all activities in support of the Air Force briefing process. The main concept behind the briefing process is to resolve issues before they become critical by presenting cost and schedule information to those in higher decision positions.

COST ESTIMATING

We have been addressing the unique and interrelated functions in acquisition management throughout this text. One function of acquisition management that gets a lot of public attention is military acquisition costs. You cannot escape the attention given to military costs in most trade journals, major newspapers, and other media. Suffice it to say that military performance is tied directly to cost effectiveness. The peacetime measurement of "how are we doing" has been attached to costs, or more exactly, cost estimates.

Cost estimating is the basis upon which budgets are formulated and revised. A program's cost estimate, therefore, is the baseline or "anchor" that keeps the program's costs tied to a number representing the program's key elements with corresponding dollar values. When all these dollars are added into the Department of Defense's (DoD) budget, or Total Obligation Authority, it takes up a sizable portion of our nation's tax dollars. It is no wonder, then, that so much attention is given to the DoD budget and to the cost estimating process. The acquisition process requires and focuses on the cost estimate of a system and the availability of adequate funding levels at the proper time. It is mandatory that cost estimates accurately reflect program financial requirements. A program's viability can be seriously impacted if measured against a less than competent estimate.

Any discussion of cost estimating usually begins with a definition. Cost estimating is defined in AFSCM 173-1, Cost Analysis Procedures, as:

> The process of projecting financial requirements to accomplish a specified objective. It includes selecting estimating structures; collecting, evaluating, and applying data; choosing and applying estimating methods; and providing full documentation.

Additionally, it should be emphasized that the cost estimating process occurs at a specific point in time and, therefore, is as accurate as the information used in its formation. The key to understanding the cost estimating process is an appreciation of its iterative and dynamic nature.

One program may go through multiple cost estimating projects, dependent on the acquisition or budget cycle. Cost estimates, therefore, serve several functions. One is to provide key information early in a program's life in the establishment of a program baseline. At this point in a program's life cycle, it is crucial to develop as accurate a cost estimate as possible because many complex and important decisions are made, based on this estimate, that affect the program throughout its life cycle until disposal.

Types of Cost Estimates: The comptroller on the product center staff, serves as an independent check of the SPO cost estimate. These cost estimates are required for development and representation of program costs in the budgetary and acquisition cycles. Cost estimates are commonly required for each milestone decision, for impact studies when the program's technical content changes, if previous cost estimates are obsolete, and upon direction or tasking to do so. Cost estimates are also used in source selections to provide support information for comparison with contractor proposals.

The type of cost estimate used depends on the requirement, program information available, phase in the acquisition cycle, and time allowed for estimate development. Generally, there are three categories of cost reviews that the comptroller performs using Program Control representatives: the component cost analysis (CCA) required on all major programs, the independent cost study (ICS), and the independent sufficiency review (ISR).

Component Cost Analysis (CCA): Air Force Cost and Accounting Agency (AFCAA) HQ USAF/ACC manages the CCA program based on AFR

173-1, The Air Force Cost Analysis Program. A CCA is a complex effort requiring a life cycle cost estimate for the Defense Acquisition Board (DAB) and/or the Air Force Systems Acquisition Review Council (AFSARC). CCAs also include a detailed risk assessment and sensitivity analysis.

Independent Cost Study (ICS): The ICS is unique in that it can be generated in the program office or by the comptroller depending on who tasked the requirement for the cost estimate. The ICS has less scope than an ICA since there is no life cycle cost estimate requirement.

Independent Sufficiency Review (ISR): The ISR is used to analyze the SPO cost estimate for completeness, reasonableness, consistency, and documentation. The key is the independent nature of the ISR as a check on the accuracy of the SPO estimate.

COST ESTIMATE CHARACTERISTICS

Understanding the characteristics on an acceptable cost estimate helps in presenting program information efficiently and effectively.

Completeness: The first characteristic is completeness. A cost estimate should include everything that the SPO has been tasked to do and deliver. Specifically, all costs must be inflated correctly and included in the estimate. These costs must conform to the WBS format and be based on actual, current information. Efforts must be made to ensure that costs included in the estimate are pertinent to the program.

Reasonableness: The second characteristic is reasonableness. The cost estimate methodology and logic should agree with current policies and should be appropriate for the program being estimated. Any data bases used as references in the cost estimate should be valid and applicable. The assumptions, learning curve, and other factors used in the estimate should be reasonable and applicable.

Consistency: The third characteristic of a good cost estimate is consistency. Any inconsistencies between the cost estimate and acquisition strategy are analyzed. The SPO's acquisition strategy or plan should be consistent with the latest program direction and schedule. All ground rules and assumptions made in the cost estimate should be consistent with current direction. Any differences between the cost estimate and previous estimates are addressed and analyzed.

Good Documentation: The last characteristic of a good estimate is good documentation. Documentation must be clear, concise, and explicit in supporting statements made in the cost estimate. Documentation must be presented and organized in accordance with AFR 173-1. It clearly explains how the estimate was developed, what methods were used in each WBS area, what models were used to generate costs and the projected accuracy of the model's estimated costs pertaining to your program, the team members involved who developed the cost estimate, and the logic behind the estimate's creation and organization. Documentation is the key to separating a good estimate from a bad one for the single reason that the estimate can be reviewed, updated, or analyzed by others separate from the program or comptroller. Because of the iterative nature of the acquisition process, one program can easily go through multiple cost estimates in its life time. A common thread needs to flow from one cost estimate to the next, if possible.

Documentation facilitates this flow, especially when the original cost estimator on that program probably will not be the same estimator on that program 10 or even 5 years down the road. Documentation is a valuable resource in cost estimating because it provides a means of transferring logic and thought across time, as a program is being developed, for showing the logic behind important decisions or reasons for historical cost growth.

8.3   Historical Reasons for Military Cost Growth

Since the public is sensitive to the cost growth of military systems, we need to have an appreciation for the basic reasons that drive this cost growth. One factor that has led to weapon system cost growth is the loss of estimating expertise continuity. Failure to analyze program costs the same way or with the same objectives or background leads to inconsistencies in cost comparisons. Documentation can help alleviate this problem.

A second factor is the use of historical costs for gauging future program costs. Is it correct to assume that current programs under development will follow cost patterns of historical weapon systems? Most people in the cost analysis field agree that historical costs are usable but with certain constraints. Adjustment factors usually used tailor program costs by comparing systems and estimated costs with historical program costs. These comparisons are useful and necessary in providing the basis on which to estimate and present costs. Subsequently, new cost estimating methods are constantly being developed in order to refine and improve the use of historical costs.

Other cost growth factors include limited program definition, difficulty quantifying program risk, limited funding, inflation, questionable material availability, and the overall failure to include all costs. Trying to account for any of these factors when developing a program's estimated cost is best left up to people with the corporate knowledge and expertise in doing cost estimates, usually Program Control and comptroller personnel.

8.4   Scheduling Tools

There are numerous methods of devising and tracking program schedules, with two of the most common being Milestone charts and Network, or PERT, charts. Tracking program schedules has great importance because of their relationship to the Acquisition Program Baseline (APB).

Milestone Charts: Milestone charts are the type of charts you see frequently throughout the program office, and an example can be found in Fig. 8.1. Typically, we find various activities along the vertical axis, and time, normally lined out in months, on the horizontal axis. Milestone charts also indicate when activities are supposed to start and end, whether or not there has been a schedule slip, and when some of the major program milestones are to occur.

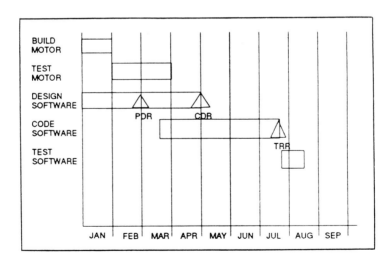

Fig. 8.1: Milestone chart

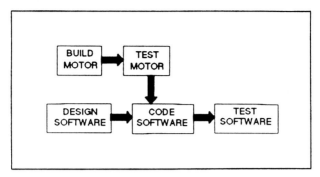

Fig. 8.2: Network chart (PERT)

The greatest advantage of this technique is that you can find concise information at a glance; milestone charts are easy to interpret. There are typically two disadvantages associated with milestone charts. The major disadvantage is that the don't show the relationships between activities, so there is no way of knowing if a slip in one activity will affect another. The second disadvantage is that the are time consuming to keep current.

Network Charts: Network charts, shown in Fig. 8.2, were first developed in the late 1950s on the Polaris program. There are basically two types: (1) Program Evaluation Review Technique (PERT), and (2) Critical Path Method (CPM).* The differences between the two are minor, and many people use the terms interchangeably. Furthermore, over the years the features of each have been added to the other with the terminology PERT used in the defense industry, and CPM in the construction industry.

The biggest advantage to this type of chart is that it shows the inter-relationships of various tasks and allows you to calculate the critical path, which gives you the minimum time a program can be accomplished in. They also act as a tool to determine where you might want to reassign resources to remove the bottlenecks from the system. The examples we have in this text illustrate this principle. Looking at the milestone chart, it would appear that software tasks would be able to continue on with no impact if the rocket motor were to have problems. The PERT chart indicates a different story, showing that rocket motor test results factor into the software coding, thereby creating a problem to software development if rocket motor problems were to develop.

---

*PERT was created by Booz, Allen, and Hamilton (management consultants) and Lockheed Aircraft Corporation, and CPM was developed by the DuPont Company for chemical plant construction.

The biggest disadvantage with a Network chart is the time involved in initially laying it out. This has been aided somewhat by the advent of software programs that will do network analysis, although the inputs to these programs, in terms of length of activities and interrelationships may be difficult to obtain. It is also a lot of work to keep this type of chart current. Both types of charts have their place in the program office.

8.5   Cost/Schedule Control System Criteria

Cost/Schedule Control System Criteria (C/SCSC) is a set of criteria used to gauge the adequacy of the contractor's financial reporting system to ensure that the government gets the cost reporting information it needs.

There are several reports that come out of the C/SCS criteria, most of which were mentioned in the Program Evaluation area. The first is the Cost Performance Report (CPR), required on major programs, which is a monthly report from the contractor on program costs and schedule. There are five different formats to the CPR, each containing different information.

Two other products are the quarterly Contract Funds Status Report (CFSR), in which the contractor provides information about funding data to aid in planning and forecasting contract fund requirements, and the Cost/Schedule Status Report (C/SSR), which the contractor submits monthly, that is similar to the CPR, but does not have as much detail and is used on smaller programs.

TERMINOLOGY

There are several terms that are important to know to analyze the data in the CPR. First of all, it is important to know that cost activities rely heavily on the work breakdown structure. The contractor takes the work breakdown structure that we give him, and then continues to decompose it to lower levels until he can match a task to a single department, forming a work package. It is through the use of these work packages that we get the information we need for the C/SCS reporting.

Budgeted Cost of Work Scheduled: The Budgeted Cost of Work Scheduled (BCWS) is the first element, and represents all of the work packages that were scheduled for completion in a given time period. The BCWS forms what is known as the Performance Measurement Baseline (PMB), which is the standard used to gauge performance against.

Budgeted Cost of Work Performed: The Budgeted Cost of Work Performed (BCWP) is the sum of the budgets for all of the work packages completed, and portions of work packages in process, at a particular point in time. This element is also known as "earned value."

Actual Cost of Work Performed: The Actual Cost of Work Performed (ACWP) is the sum of the actual costs for all of the work packages completed, and portions of work packages in process, at a particular point in time.

Estimate at Completion: The Estimate at Completion (EAC) is the estimate of what the program will actually cost when it is finished. The initial Estimate at Completion (EAC) is derived by adding all of the budgets of the work packages together, and then adding in any management reserve that the contractor has, if it appears that he will use it. There are many techniques for calculating EAC, each with attendant strengths and weaknesses, all of which are out of the scope of this text

Data Analysis: Now that we know what the different type of data are, let's look at how they are normally displayed and analyzed. Fig. 8.3 shows an example of what typical Cost Performance Reports (CPRs) look like graphically. We can see all of the terms we just discussed shown in this figure. The first thing to notice is that the Y-axis represents cost, which is cumulative over the program, and that the X-axis represents time.

COST PERFORMANCE REPORT

The Cost Performance Report (CPR) provides us with a check on the program at any point in time. This information is normally sent to the SPO in a tabular format, although it can readily be converted to the graphical format shown here. Although we have shown all the lines out to program completion in Fig. 8.3, it is not too hard to imagine the ACWP and BCWP stopping at the point marked "current time." You will probably notice that BCWS and BCWP are equal at the end of the contract, something that occurs by definition, since they both work from the same budgeted numbers.

As it can be seen from Fig. 8.3, the difference between BCWP and BCWS provides a schedule variance in terms of cost, and the difference between the BCWP and the ACWP indicates a cost variance. These variances can be both positive and negative, and both bear watching. A negative number from either of these relationships indicates something unfavorable that needs checking.

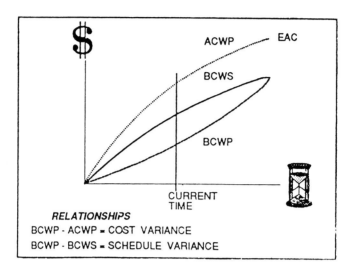

Fig. 8.3: Typical CPR data (graphical format)

There are a number of possible explanations for favorable cost variance, and additional information will have to be obtained from the contractor to help determine the cause of the variance. Some of the possible explanations are: poor initial planning and estimating, a technical breakthrough, the cost of material or labor is lower than planned, the contract is front end loaded, or the method of earning BCWP is inappropriate.

There are also a number of reasons for unfavorable variances, some of which are: poor initial planning or estimating, technical problems, the cost of labor or material is higher than planned, inflation, new labor contracts, work stoppage, and contract requirements changes.

References

[1] AFSCP 800-3, "A Guide for Program Management"

[2] AFR 800-25, "Acquisition Program Baselining"

[3] AFR 173-1, "The Air Force Cost Analysis Program"

Chapter 9

CONTRACTING MANAGEMENT

9.1 The Federal Contracting Process

INTRODUCTION

Now that we have presented a basic background in the acquisition life cycle process, we will turn our attention to the means of getting a contractor to perform work for us, i.e the contracting process. We have already discussed that we can go through the solicitation process at each phase in the system life cycle, although we have not provided the details of the process. Contracting is a complex and a labor intensive process. It is for these two reasons that we need to understand the contracting process and how we can better make it work for us. Figure 9.1 is a flow diagram describing the federal contracting process, which is discussed in this chapter.

PURPOSE

The purpose of the contracting process is two-fold. The first purpose of the contracting process is to insure that all interested contractors have an opportunity to bid for work, and to have that bid be comprehensively, impartially, and equitably reviewed for consideration. The second purpose is to determine a source that will best meet the government needs at the best value.

PUBLIC LAW 98-369

In 1984 Congress passed Public Law 98-369, The Competition in Contracting Act (CICA). This legislation made sweeping changes in many areas of contracting, emphasizing competition as a key ingredient of the process. In the contracting area, formal advertising was renamed the sealed bidding process and negotiation renamed the competitive proposal process. Both processes were acknowledged as equally legal and effective methods, given the circumstances of the acquisition. No longer do contracting officers have

151

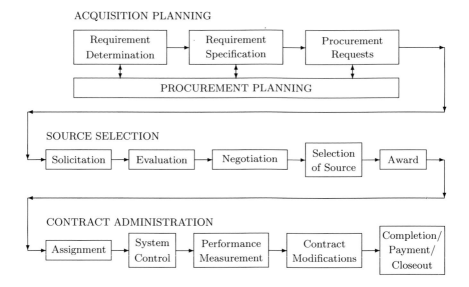

Fig. 9.1: Federal contracting process.

to process a waiver in order to use negotiation. However, approvals must now be obtained to use less than full and open competition in an acquisition.

TYPES OF SOLICITATION

The method of solicitation generally takes one of two major forms, either sealed bids, or competitive proposals. If sealed bids (formerly called "formal advertising") are used, the solicitation instrument is called the Invitation For Bids (IFB).

Sealed Bids: The sealed bids method is quite rigid and inflexible in its implementation. Sealed bids has its origin in a congressional enactment of 1809 which stipulated that formal advertising would be required in the procurement of all supplies and services required by the Government when the situation is practicable and feasible. The original purpose was to preclude granting special consideration to any individual or groups of individuals and to afford the Government the benefits of open competition. In the years to follow, there have been departures from this stringent method of procurement, particularly during periods of war, national emergencies, or occasions wherein resort to formal advertising was inappropriate. Notwithstanding, it remains to this day the method preferred by Congress and prescribed by law when the situation is practicable and feasible.

In this respect, the Federal Acquisition Regulation (FAR) defines the situation as being practicable and feasible if it meets all of the following prerequisites:

1. *Specifications:* To assure free and full competition there must be detailed specifications so that all potential bidders understand the requirement of the particular transaction. The specifications must be free of ambiguities, firm and not susceptible to unnecessary and frequent change, and include restrictive conditions only to the extent necessary to satisfy the minimum needs of the government or as required by law.

2. *Adequate competition:* There must be an adequate source of supply and a sufficient number of suppliers who indicate an interest in obtaining the contract.

3. *Adequate time:* There must be ample lead time from receipt of the program directive to the delivery date of the supplies or services.

4. *Price is Adequate Criterion:* Price and price related criteria constitute adequate contract award criteria. If technical discussion is required, the sealed bid method cannot be used. Also, an IFB must result in a Firm Fixed-Price-type contract only. In periods of high inflation, a Fixed-Price contract, with a Economic Price Adjustment (EPA) clause, also may be used.

Sealed Bidding Process: First, the specification and other aspects of the requirement are reviewed and the necessary planning accomplished. Next, the Invitation for Bids (IFB) is developed. The IFB indicates to potential bidders what the requirement is, how the resulting contract will be structured, and what steps must be followed to properly bid on the requirement. The IFB is mailed to the bidders on the Bidders' Mailing List. From the date of mailing the IFB, a period of at least 30 days or so is allowed for the contractor to review the requirement and to prepare and submit a bid. As bids are received, they are placed unopened in a locked bid box. On the day and at the time specified in the IFB, bids are publicly opened and read aloud, with as many of the bidders present as desire to be. All those in attendance can see the range of bids and who submitted the apparent low bid and price related criteria.

Once an apparent low bidder is identified, it must be determined if the contractor is responsive to the requirements of the IFB and responsible in terms of financial, managerial, and technical capabilities. To be determined responsive, a contractor submits a bid that complies in all material respects with the Invitation for Bids. To be determined responsible, a contractor must demonstrate adequate financial resources to perform the contract and have a satisfactory record of performance, integrity and business ethics. If the low-bidder is determined to be both responsive and responsible, the contract is signed by the contracting officer and awarded to the low bidder with no discussions allowed. If a review of the bid forms or checking the responsiveness and responsibility items reveals a problem, the apparently low bidder may be rejected from the competition. This is done in strict accordance with specific rules, and the award is made to the next lowest bidder who is both responsive and responsible.

This method of contracting has much appeal to contractors because it is used in many commercial acquisitions, is inherently fair to all competing bidders, and highlights efficiency in doing the work. However, as requirements increase in complexity, the circumstances favoring sealed bidding are not always present. In acquisitions of complex major weapons systems, there is seldom a complete and definitive specification of the work to be done or even an idea of what the finished product should look like. Therefore, detailed technical discussions are needed which are not allowed under sealed bidding. Additionally, there are often instances where only a single contractor could, or would, perform the contract, thereby eliminating the possibility of competition. Because of these developments, competitive proposals (negotiation) has become the prevalent method used, despite a regulatory preference for sealed bids.

The sealed bid method has relatively little application in the systems contracting arena; it is used primarily in base-level contracting and in some centralized spare-parts acquisitions.

Competitive Proposals: Competitive proposals (formerly called negotiation) involves sending Requests for Proposal (RFP), receiving and analyzing the proposals, discussing technical requirements, and negotiating price, delivery schedule, terms, and conditions of the contract. In systems contracting, the competitive proposals method (still referred to as negotiation) is used almost exclusively.

While the RFP's content is substantially the same as an IFB, it does differ in certain respects. For instance, usually the RFP is written, but in certain situations, such as an emergency or other unusual circumstance, it may be issued by telephone or telegraph. While specifications and/or drawings are preferable, they are not mandatory, as is the case in formal advertising. However, under normal conditions where the competitive proposals method is utilized, the contracting officer preparing the RFP must conform to all the requirements prescribed for the IFB process (i.e., adequate distribution to qualified offerors and circulation of the RFP well in advance of the time set for the closing of the offer period).

Competitive Proposals Process: The request for competitive proposal process can be a long one, with a number of various documents associated with it. We will examine this process in greater detail than we did the sealed bid process, because it is far more complex and used for all major system acquisitions. Once a Program Management Directive has been released, the program office, or cadre of a newly forming program office will start the planning that goes into the request for proposal solicitation process.

Acquisition Strategy: Although there may be some initial steps, like setting up a technical library, a bulletin board (physical and/or electronic), or performing a request for information from industry, the acquisition strategy is the first formal step in starting the competitive proposal process. The Acquisition Strategy is an overall strategy for ALL of the life cycle phases that includes information on management, technical, resource, procurement and contracting, testing, training, deployment, support, and other aspects critical to the success of the program. The Acquisition Strategy is more fully discussed in DoD Instruction (DoDI) 5000.2.

Acquisition Strategy Panel: The Acquisition Strategy Panel (ASP) is a group of senior acquisition executives, whose grades are commensurate with the dollar value of the program. The purpose of the Acquisition Strategy Panel is to review the Acquisition Strategy, and provide written feedback to the program manager, which then must be addressed in the Acquisition Plan. One of the agenda items for the Acquisition Strategy Panel is to recommend a Source Selection Authority, the individual who will make the final decision in selecting the successful offeror.

Acquisition Plan: The Acquisition Plan serves as a top level planning document to insure the effective integration of the various acquisition events, documents, and activities in any given phase of the program. The Acquisition Plan serves a threefold purpose:

1. Ensure that the Government meets its needs in the most effective, economical, and timely manner.

2. To reduce acquisition risk by causing the acquisition planner to think through the acquisition process before the fact so that he or she is

aware of the steps to be taken, activities to be integrated, problems to be resolved, and risks to be expected.

3. To execute an effective integration of the various functional plans such as the System Engineering Master Plan, Integrated Logistic Support Plan, Test and Evaluation Master Plan, etc.

The Acquisition Plan serves as a "living document" in that it "matures" or "evolves" over time as more current information becomes available and relevant. Because of the evolutionary nature of the process, the FAR requires that the Acquisition Plan be reviewed and revised throughout the acquisition cycle. The Procuring Contracting Officer (PCO) is responsible for ensuring the Acquisition Plan is written and maintained.

Source Selection Plan: The Source Selection Plan (SSP) is a major document prepared early in the solicitation process, prior to RFP release, that provides the ground work for subsequent selection of a contractor to do work for the Government. The SSP usually provides for a smooth, efficient source selection process in competitive solicitations. The SSP establishes procedures for accomplishing three prime objectives:

1. Ensure impartial, equitable, and comprehensive evaluation of competitor's proposals and related capabilities.

2. Maximize efficiency and minimize complexity of solicitation, evaluation, and the selection decision.

3. Select the source whose proposal has the highest degree of realism and credibility and whose performance is expected to best meet Government objectives at the best value.

The Source Selection Plan is written by the Program Manager and approved by the Source Selection Authority (SSA).

REQUEST FOR PROPOSAL PACKAGE

The Request for Proposal (RFP) Package is also being prepared at the same time the Acquisition Plan and Source Selection Plan are being prepared. The Request for Proposal (RFP) Package is prepared in accordance with the Uniform Contract Format (UCF). The UCF is divided into four parts, with a total of 13 sections.

Part I: The first part of the RFP contains Sections A-H and is known as "The Schedule."

Section A: Section A is the Solicitation/Contract Form, which contains information like the RFP number, proposal due date, government points of contact, table of contents, and other pertinent information.

Section B: Section B is the Supplies or Services and Prices/Cost, and contains a brief description of the supplies and services and quantities required, the unit prices, and total prices. This is the section of the RFP where Contract Line Item Numbers (CLINs) will be found.

Section C: Section C is the Description/Specification/Work Statement, and contains a more elaborate description of the items contained in Section B, and describes what is to be accomplished, but does not describe how the tasks are to be performed. The Statement of Work can be found in this section, although it is typically found in Section J.

Section D: Section D is Packaging and Marking, and contains information on requirements for packaging and marking of items to be delivered.

Section E: Section E is Inspection and Acceptance, and contains information on how the government will inspect and conditions for acceptance of items to be delivered under the contract.

Section F: Section F is the Deliveries or Performance Section, and specifies the requirements for time, place, and method of delivery or performance for items to be delivered under the contract.

Section G: Section G is Contract Administrative Data, and contains accounting and appropriations data and required contract administration information and instructions.

Section H: Section H is Special Contracts Requirements (Special Clauses), and contains contractual requirements that are not included in other parts of the contract, including specialty clauses that only pertain to that particular acquisition. This section is where information on things like data rights would be contained.

Part II: Part II is known as Contract Clauses and contains only *Section I*, which is known as General Contract Clauses. This section contains all of the clauses required by regulation or law, and is referred to as the boiler plate.

Part III: Part III is the List of Documents, Exhibits and Other Attachments, and contains only Section J. This section contains or lists documents, attachments, or exhibits that are a material part of the contract. Examples of some of the items found in this section are the work breakdown structure (WBS), the Statement of Work (SOW), the Contract Data Requirements List (CDRL), the Lists of Specifications, List of Government Furnished Property (GFP), and others.

Part IV: Part IV is Representations and Instructions, and contains the last three sections (K, L, and M) of the uniform contract.

Section K: Section K is Representations, Certifications, and Other Statements of Offerors or Quoters, and contains required contractor representations, certifications, and other information required from each offeror.

Section L: Section L is Instructions, Conditions, and Notices to Offerors and Quoters, and tells the offerors what is to be provided in their proposal as well as how it should be formatted. It guides offerors in preparing their proposals, and emphasizes any government special interest items or constraints.

Section M: Section M sets forth the Evaluation Factors for Award, which form the basis for evaluating each offeror's proposal. It informs offerors of the relative order of importance of assigned criteria so that an integrated assessment can be made of each offeror's proposal.

After the contract is awarded, the first three parts of the Request For Proposal form the Contract, which also may have references to *Section K*, which is then signed by the government and the winning contractor.

Once the RFP has been pulled together in a releasable draft format, the requirement is advertised in the Commerce Business Daily (CBD), and copies of the draft RFP are sent to interested industry companies for review and comments. Once the comments have been received and evaluated, the prospective offerors are given the disposition of their comments, and the valid comments are rolled into the final RFP package.

When the package has been updated, it is sent to the Solicitation Review Board (Murder Board) for review. The "Murder Board" is made up of more senior acquisition personnel, and provides a comprehensive review of the acquisition approach and its implementation. Once the "Murder Board" comments have been incorporated, the RFP is reviewed by the legal office, another round of advertising is initiated in the CBD, and the

RFP are distributed.

Source Selection: At some point in time prior to receiving the proposals, the Source Selection team drafts Evaluation Standards. These Evaluation Standards should reflect the criteria given to the contractor in Section M of the RFP. The Evaluation Standards are normally based on the Statement of Work (SOW), and the Specification, and will be used to evaluate each offeror's proposal.

Once the proposals are received, the source selection process begins. Proposals are compared against the evaluation standards, and ratings and a narrative are provided for each of the evaluation standards. A Government team may go to offeror's facilities to further evaluate the offeror's capacity and capability to perform the requested work.

Clarification Request (CR): Inevitably, there will be ambiguities or discrepancies in the contractor's proposal. To address these, clarification request will be sent out to the contractor after initial technical evaluation to permit the contractor to justify their position or clear up misunderstandings. Generally the contractor has only a few days to respond to the clarification request, which helps to keep source selections on schedule.

Deficiency Report (DR): A Deficiency report is more severe than a clarification request, and is only issued if there is a major discrepancy that would cause the contractor to be rated unsatisfactorily in a particular area of evaluation. We give the contractor the opportunity to correct the deficiency because we want to maintain as many viable offerors as possible in the competitive range.

Both of these requests (CRs and DRs) flow through the Procuring Contracting Officer (PCO), who insures that information is not inadvertently released that might give one contractor an unfair advantage over another.

After the proposals have been initially evaluated, a mid-term briefing is prepared for the Source Selection Authority (SSA). The SSA determines the offerors that have a viable chance of award. This group is called the competitive range. These offerors are called in for discussions to receive any final proposal clarifications. After the meetings, the offerors prepare a Best and Final Offer (BAFO), final evaluations are submitted, and a recommendation is passed to the Source Selection Authority (SSA) for a final decision. Once the SSA has made a decision, the proper government personnel are notified, the appropriate Congressional personnel are notified, and the announcement is made, concluding the solicitation process.

SOURCE SELECTION PERSONNEL

We have discussed several of the individuals and groups involved in the Source Selection process, but haven't provided a description of them yet, something we rectify in this section.

Source Selection Authority: Prior to the issuance of an RFP, a Source Selection Plan (SSP) is approved by the Source Selection Authority (SSA), the Government official in charge of selecting the source. The program manager is responsible for drafting the plan and obtaining its approval from the SSA. The SSP complements the Acquisition Plan and summarizes the overall acquisition strategy contemplated for the program. The SSP includes a discussion of the extent of competition contemplated, a description of the evaluation techniques to be used, and the schedule for significant actions required between the designation of the SSA and signing of the definitive contract. Also included in the SSP is a description of the organizational structure to be used in the source selection process. The organization is normally composed of three levels: the SSA, the Source Selection Advisory Council (SSAC), and the Source Selection Evaluation Board (SSEB). The Source Selection Authority is chosen based on the size of the program, and may be somebody as high as the Defense Acquisition Executive, or as low as the Procuring Contracting Officer (PCO).

Source Selection Advisory Council: The Source Selection Advisory Council (SSAC) is a group of senior military and/or Government civilian personnel designated to serve as the staff and advisor to the SSA during the process. The SSAC reviews the SSEB findings, prepares a proposal analysis of each offer, and compares the proposals to one another. The SSAC is the body that considers the contractor's past performance.

Source Selection Evaluation Board: The Source Selection Evaluation Board (SSEB) is a group of military and/or civilian personnel, appointed by the SSAC, representing various functional and technical disciplines. Their task is to evaluate proposals against established criteria, not proposals versus proposal, and to develop summary facts and findings during the source selection process. The SSEB is the heart of the selection team. Its membership typically includes personnel from logistics, cost analysis, operational, contract, legal, and technical areas, and is typically composed of personnel that will work in the program office after the contract has been awarded.

The SSEB oversees the source selection team that performs the proposal evaluation. You can look at the evaluation team as the workers and the evaluation board as the managers from each functional discipline. The analogy stops here, though, because the evaluation board members are just as involved in the evaluation as the team members. The evaluation board members are the individuals that will decide what clarification requests and deficiency notices will be forwarded to the offerors. The SSEB also prepares all the briefings and analyses for the SSA.

CONTRACT TYPES

Once a source is selected, the next step is to prepare a valid contract established for work to begin. The complexity of Government acquisition has led to development of a wide variety of contract structures, as indicated by the compensation arrangement given the contractor. In general, there are only two basic types of contracts, fixed-price or cost reimbursement, but each type may have any of several "variations on a theme."

Fixed-price: Fixed-price contracts provide for a firm price or for a price which is adjusted only within fixed limits. This places the maximum cost risk on the contractor who is in the position to control the costs. When conditions are appropriate for its use, the firm fixed-price (FFP) type contract is preferred. Before work is started, an agreement is reached on the price to be paid for the completed job; thereafter, no adjustments are made to the price regardless of actual cost experience. Thus, if final costs are less than agreed-to price, there is a profit; if more, there is a loss. This type of contract is required for sealed bid procurement, and may be used for competitive proposals. Fixed price contracts are no longer allowed for development contracts that exceed $10 Million unless the Defense Acquisition Executive (DAE) approves.

Cost-reimbursement: At the opposite end of the contract-type continuum from firm fixed-price is the extreme case of the cost reimbursement-type contracts, the cost-plus-fixed-fee (CPFF) contract structure. In it the Government promises to reimburse all valid costs incurred in pursuing the contract objective and to pay a fixed amount of fee on top of that.

Risk (of encountering technical obstacles, of overrunning targeted costs, of missing required delivery dates) is the predominant element that determines where on the contract-type continuum the appropriate structure is found.

There are two additional things that we need to look at in greater detail prior to departing this section, the Work Breakdown Structure, and the Statement of Work. We would also examine the Contract Data Requirements List in greater detail here, but it is more appropriately covered in the chapter on data. We give some special emphasis to the Work Breakdown Structure and the Statement of Work because these become valuable tools to acquisition personnel in understanding how the program is structured in terms of its products, and what we expect the contractor to do for us. It is important to have a good understanding of both of these documents to be able to understand your particular program.

WORK BREAKDOWN STRUCTURE

The common thread that ties all plans and documents together is the Work Breakdown Structure (WBS). The WBS is a product-oriented family tree composed of hardware, software, services, and data that result from project engineering efforts during the development and production of a defense materiel item or weapon system.

MIL-STD-881B describes the WBS for use by both contractors and DoD components in the development of work breakdown structures for defense material items. Most contractors manage their work according to a format similar to this, at whatever level is necessary to meet established schedules, and that has been validated by the Defense Department. The Air Force manages contractor efforts according to information presented by the contractor in WBS format.

The WBS, in its several forms, is an extremely useful device as project or program managers engage in planning and controlling their programs. MIL-STD-881B is intended to be a guide. Rigid adherence to the formats is not required. A WBS, however, if sufficiently written, defines the program's total objectives; it relates the various work efforts (parts) to the overall product (whole system). The WBS is the foundation for:

- Program and technical planning.

- Statement of Work preparation.

- Schedule definition.

- Cost estimation and budget formulation.

- Progress status reporting and problem analysis.

There are two basic WBS formats highlighted in MIL-STD-881B, the program work breakdown structure and the contract work breakdown structure. DoDI 5000.2 Part 6-B further expands these formats to a total of four work breakdown structures.

Initial Program WBS: The initial program work breakdown structure is defined in DoDI 5000.2 as the upper three levels from one or more defense materiel items from MIL-STD-881B that are tailored through deletion and addition to entirely define the program being developed. Deviations are allowed from the standard defined elements when a unique requirement, normally defined through the systems engineering process, exists, and additional elements at lower levels may be specified if necessary.

The initial program work breakdown structure is defined as the program work breakdown structure in MIL-STD-881B. This work breakdown structure is prepared by the Government in getting ready to build a request for proposal. Figure 9.2 shows a hypothetical initial program WBS for a missile system. The terminology used in the figure doesn't necessarily coincide with the terminology in MIL-STD-881B. You will notice that the missile system is at the top of the WBS, or at level 1. With each lower level of the WBS, we continue to define smaller elements of the system and show how the elements are interrelated. Work breakdown structures continue to go down to as many levels as the contractor feels is appropriate to define and manage his work packages, although we typically only get reporting information to level 3.

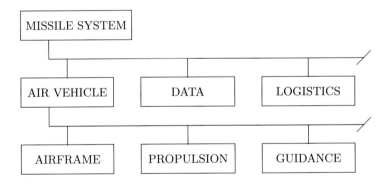

Fig. 9.2: Initial program WBS for a missile system.

Preliminary Contract WBS: Once the initial program work breakdown structure is complete, preliminary contract work breakdown structures, as defined in DoDI 5000.2, are developed for each contractor. This is necessary where different contractors will be performing different portions of the work.

A good example of this is an aircraft acquisition. Typically, although the engines and their lower level breakouts are defined as a part of the initial program work breakdown structure, they are normally developed under a separate contract and provided to the prime contractor for the aircraft as government furnished equipment. This would provide different preliminary contract work breakdown structures for the prime contractor for the aircraft and for the engine contractor.

This is the work breakdown structure that will be included in a request for proposal package, and there will be only one preliminary contract work breakdown structure for each request for proposal. This work breakdown structure is also developed by the Government.

Contract WBS: The contract work breakdown structure is defined in both DoDI 5000.2 and MIL-STD-881B as the work breakdown structure that is extended to lower levels by the contractor to use in structuring his program and to provide management information back to the Government.

The contractor will extend the work breakdown structure down to whatever level he finds necessary to manage the work and to define work packages, which are discrete portions of the project that can be costed to a single organization and act as the basic building block for program management and cost reporting.

Final Program WBS: The final program work breakdown structure is defined in DoDI 5000.2 as the work breakdown structure that compiles all of the elements of the contract work breakdown structure(s) with the initial program work breakdown structure.

Work breakdown structures are defined by using one or more of the defense materiel items found in MIL-STD-881B. The defense materiel items defined in the standard are: (1) Aircraft/Helicopter systems, (2) Electronic systems, (3) Missile systems, (4) Ordnance systems, (5) Ship systems, (6) Space systems, and (7) Surface Vehicle systems. It may be necessary to use elements from several of the materiel items in defining a system.

A good example of this is the ground launched cruise missile system, where we would find elements from both the missile and surface vehicle materiel items. Work breakdown structures can be developed for either entire systems, or parts of systems that are managed as an entity like the jet engine discussed above, by applying the various materiel items at the appropriate level.

Work breakdown structures must also correlate with the specification tree, contract line items, configuration items, data items, and statement of work tasks, and this correlation must be confirmed in checking the contract work breakdown structure.

Other considerations for the work breakdown structure are contained in DoDI 5000.2. When building a WBS we must insure that Integrated Logistics Support is defined at an appropriate level, that software will be identified with the hardware that it supports, that overall system software that interfaces with more than one equipment item will be called out at the appropriate level, and that functional elements like engineering, tooling, quality control, and manufacturing are not represented as work breakdown structure elements.

STATEMENT OF WORK

With the WBS defined and developed, it is possible to prepare the Statement of Work (SOW). The SOW is a document that defines efforts to be accomplished ranging from small research studies to the acquisition of a major weapon system. It establishes non-specification tasks/requirements and identifies the work effort. The SOW is a tasking document that defines the scope or outer limits of the contractor's effort. The SOW is part of the RFP and the contract.

There are five types of SOWs, one for each of the Acquisition Phases 0 – IV, and one for non-personal services. Within each of these SOWs there are three major sections.

Section 1 – Scope: This section defines the overall purpose of the program and what the Statement of Work applies to.

Section 2 – Applicable Documents: This section lists the various specifications and standards that are referenced in the SOW. It is typical that we have the contractor perform a task in accordance with (IAW) a particular standard, such as DoD-STD-2167 for software development. Specifications and standards convey requirements to the contractor.

Section 3 – Requirements: This is the section where the actual tasking appears. The tasks in this section should track with the work breakdown structure.

The SOW is a key element of the RFP, usually an attachment in Part III, and serves as a basis for contractor response and subsequent Government evaluation of proposals in source selection. After contract award, requirements of the work statement (and associated specifications) constitute the standard and discipline for the contractor's effort. It comprises the baseline against which progress and subsequent contractual changes are measured. Both the Government and the contractor look to the SOW as a key document defining the responsibilities of both parties.

There are several considerations to be considered in developing a Statement of Work. First of all, the Statement of Work (SOW) should track to the WBS to insure that all WBS elements have been tasked to the contractor. This acts as a cross-check of completeness. The SOW should not task the contractor to deliver data. That is something that is done in the contract data requirements list. The SOW should task the contractor to perform some tasking, and if that tasking generates data, it should be annotated at the end of the SOW paragraph with the Contract Data Requirements List (CDRL) sequence number and the Data Item Description (DID) associated with it.

When we want the contractor to perform a task, we should use the word "SHALL" for that tasking. It is possible to use the word "will" instead of "shall", but there is absolutely no question in anyone's mind if we only use "shall." We should not try and get fancy with our SOW paragraphs; we need to be very clear and concise. It is always safe to start each paragraph with "The contractor shall..." Finally, we want to insure that no performance requirements show up in the SOW, these belong in the Specification. The problem here is that the SOW and Specification are many times developed in parallel, and we may have conflicts between the two documents; conflicts that are best resolved by leaving performance requirements in the Specification.

SUMMARY

The RFP is the vehicle the Government uses to ask industry for proposals for a system using the competitive proposal procurement method and is one of the most important documents in the acquisition cycle. All of the preparation and planning for an acquisition goes into the RFP as the key communication to potential contractors on exactly what, how, and when the Government needs to buy. When the government issues an RFP, it is describing its needs for particular goods or services, and is soliciting, from industry, proposals to fulfill those needs. Competitors (or offerors) submit proposals (or offers) in response.

The Government conducts a source selection, negotiates with the winner, and the contracting officer signs (accepts) to form a binding contract. The RFP has particular significance in this process in that the clarity and coherence with which it is constructed can dramatically affect the events that follow–favorably or unfavorably. For instance, how well the Government clearly communicates its needs in the RFP certainly influences the quality of proposals received, ease of difficulty of conducting source selection and negotiation and, ultimately, relative success or failure of contract performance.

9.2   Contract Administration

INTRODUCTION

Contract management is an important function that program management personnel need to understand. The contract is the key instrument that delineates responsibilities of both parties to the contract–industry and Government. This chapter presents some basic terminology, processes, regulations, and guidelines the Government uses in managing contracts. The major contracting offices and their roles and responsibilities are delineated in this chapter. Administering Government contracts is highlighted briefly. A constant concept that extends throughout the contract management process is that, contracting is a complex and time-consuming process designed to ensure fair and equitable treatment for industry contractors while ensuring that Government interests are achieved.

CONTRACT MANAGEMENT

The social need for legally enforceable contracts is obvious. The efficient conduct of business in the absence of enforceable rights and obligations would be impossible. This section presents the six elements that must be included in a contract.

Contracting is the service function designated to actually effect the acquisition process. That is, it provides the legal interface between the buying agency, for example the Air Force, and the selling organization, usually a civilian contractor. All the terms of the agreement between the buyer and seller must be reduced to writing and structured according to law and regulation in order to protect the rights of both parties. The resulting document is a contract which binds both parties to its provisions. Specifically, a contract is an agreement between two or more persons, based on a promise or mutual promises, which establishes an obligation that the law will enforce. Government contracts must conform to the principles of law which have governed the formation of contractual relationships for centuries.

Basic Elements: The basic elements required to be present in order to have a valid contract can be presented very simply, although legal cases on any part of an element can be very complex. Basically, there must be at least two parties to the contract, each of whom has the legal capacity to act for its organization. An offer must be formulated and communicated, and an acceptance must be made and manifested. When an offer or counteroffer is accepted, an agreement is created. An agreement, although basic to a contract, is not in and of itself a contract, as other ingredients must also be present. Consideration must be present, meaning that each party gives to the other something of value to guarantee its promise to perform. The terms of the agreement must be clear and certain, and the objective of the contract must be some legal act. Finally, the contract must be in the form required by law.

CONTRACTING AUTHORITY

Contracting officers, acting within the scope of their properly delegated authority, are the only people authorized to commit the Government to a contractual obligation. The authority to contract, while not specifically set out in the Constitution, has always been considered an inherent function of the Government in carrying out its duties. Both the legislative and executive branches have this inherent contracting authority which can be delegated to the level best equipped to carry out the action. Thus, contracting authority is delegated from the President down the chain of command through the Secretary of Defense and the Secretary of the Air Force to the appropriate working level where a properly appointed agent of the Government actually enters into a contract with a commercial firm in order to acquire its valid needs.

As evidence of this authority, a document called a warrant is issued to contracting officers. The warrant establishes their legal capacity to act for the Government and specifies any bounds on it. In large organizations, contracting officers may be asked to specialize in certain areas of contracting. A contracting officer appointed primarily to create and enter into new contracts is referred to as a principal or procuring contracting officer (PCO). An administrative contracting officer' s (ACO) main task is to administer previously awarded contracts, making certain that the contract provisions are carried out in performance. In relatively rare cases of specialization, a termination contracting officer (TCO) may be appointed to terminate or end contracts before performance is completed.

All contracting officers must meet high standards of knowledge, experience, business acumen, and ethical behavior, for they are the official representatives of the Government to American and foreign commercial firms.

Procuring Contracting Officer (PCO): A contracting officer appointed primarily to create and enter into new contracts is referred to as a principal or procuring contracting officer (PCO). The "P" can also stand for primary or procurement. There is such a wide variety of terms because this type of contracting officer is not spelled out specifically in the Federal Acquisition Regulation (FAR). This is the type of contracting officer we find in our

SPOs and working our source selections.

Administrative Contracting Officer (ACO): An administrative contracting officer's (ACO) main task is to administer previously awarded contracts, making certain that the contract provisions are carried out in performance. This contracting officer resides within the Defense Contract Management Command.

Termination Contracting Officer (TCO): In relatively rare cases of specialization, a termination contracting officer (TCO) may be appointed to terminate or end contracts before performance is completed.

Buyers: Although not contracting officers, buyers play a very important part in our contracting process. Warrants are very controlled items, and there are not enough contracting officers available to accomplish all of the work necessary to enter into every contract.

Buyers prepare most of the documents and files that the contracting officer will review and sign on behalf of the government. You will probably be working with a buyer if you have to work an engineering change proposal through your SPO. The buyer will assemble the evaluations, develop a negotiating position, and then negotiate a contract prior to giving the paperwork to the contracting officer for definitization. Definitization is the process of sending the final paperwork through the system to officially have a contractor start work.

CONTRACTORS: PRIMES/SUBS/ASSOCIATES

This section introduces the basic types of contractors that typically deal with DoD Government contracts. Not every weapon system development and production effort has a prime contractor, subcontractors, vendors, or associate contractors. More than likely, however, you will be exposed to them at some point in your acquisition career. Understanding your contractors and their relationships to one another and the Government is essential knowledge for the Air Force program manager.

One fact needs to be understood about contract and relationships between industry and Government: there may be more than one contractor or supplier per contract. Given the complexity of the weapon systems typically developed and produced for the Department of Defense, the chances are great that there will be multiple, if not hundreds, contractors or suppliers involved. These contractors or suppliers have different responsibilities and tasks for each contract, and for each program depending on how negotiations were conducted and ultimately reflected in the contract.

Some contractors may end up as "prime" contractors. Prime contractors are usually the principal or only contractors performing under contract. The prime contractor leads efforts to fulfill the contract as the primary partner with the Government. In this type of relationship we have something called "privity," which is a legal relationship between two parties in the same contract

Another type of contractor is the subcontractor. The subcontractor enters into a contract with a prime contractor to perform a piece of the work or fulfill a portion of the contract. In this relationship, privity exists with the prime contractor, but not with the Government, since we do have a direct contractual relationship. This lack of privity provides a barrier into the full visibility of our sub-contractors, and makes us rely on the prime contractor's permission to have discussions with the sub- contractor.

Finally, an associate contractor is usually a prime contractor working in conjunction with another prime contractor to complete the contract. A good example of this is a missile program that has to be connected to an aircraft. In this case, the missile and aircraft contractors would have an associate contractor agreement between the two of them, and we would have a memorandum of agreement (MOA) between the corresponding government offices.

CONTRACT CHANGES

Far more time is spent in contract administration than in any other single area of the contracting process. It is necessary to understand some of the terminology used in conjunction with contract administration, and to know some of the functions of the Defense Contract Management Command (DCMC).

One of the most common things we will experience on our contract, is change to the contract. The Contracting Officer is the only person authorized to obligate the Government or make changes to the contract. No other person should direct or encourage the contractor to perform work that should be the subject of a contract modification. Contractual changes happen through several methods, some better than others, and it is important that we understand some of the concepts that go along with this change process.

Change Order: A change order is a written order, signed by the contracting officer, that directs the contractor to make changes authorized by the contract. These change orders may be unilateral or bilateral. In the normal day-to-day working of a contract, we normally associate a change order as something we have directed the contractor to do without first negotiating the change. You might ask yourself why we would do something like that. Generally speaking, the only reason for issuing a unilateral change order is for expediency. There are some changes to the contract that are time critical, and in order to gain the maximum benefit, we will order the contractor to change the contract, provide a not to exceed price, and then negotiate the cost, and definitize the change later.

A unilateral change order puts the Government at some cost risk. After all, how many contractors will come back and negotiate for less money than we said we were willing to pay. At any rate, unilateral change orders are generally discouraged by higher management, and there normally has to be some compelling need to initiate one.

The far more common method that we see is the bilateral change order, where the contractor provides us information on the impending change, we gather necessary information, negotiate a price for the contract change, and then definitize the change.

Constructive Change: Constructive change is a contract change based on Government conduct, including actions and inactions, which is not a formal written change order, but has the effect of requiring the contractor to perform work different form, or in addition to, that prescribed by the contract. A constructive change can be brought about by a seemingly harmless discussion between Government and Contractor personnel if the contractor thinks that the Government is requesting him to do something that is not currently required in the contract. Once a constructive change has been claimed by the contractor, the contracting officer may or may not ratify the change to incorporate it formally into the contract.

Ideally, the contractor knows he should not do work that is not signed out by the PCO, but not all of a contractor's personnel will. Working level people will many times go out of their way to meet our needs, and may be committing the contractor to perform work that was not bid for. One method of avoiding this is to use a disclaimer statement before every meeting that says "The contractor is not to take any of the discussions in this meeting as contractual direction. Contract changes are only authorized through the contracting officer." We need to use this disclaimer statement, and be intimately familiar with the statement of work (SOW), specification, and contract data requirements list (CDRL) to help minimize the possibility of accidently initiating a constructive change.

9.3   Defense Contract Management Command (DCMC)

No discussion of contract administration would be complete without discussing the Defense Contract Management Command (DCMC). This command was established in 1991 and has its roots in the older, fragmented contractor oversight organizations that we used to have. DCMC is organized as part of the Defense Logistics Agency (DLA), and has its headquarters at Cameron Station, VA. DCMC is geographically split into five different districts in the U.S., with district headquarters in Boston, Philadelphia, Atlanta, Chicago, and Los Angeles, and also has an international organization for foreign contractors. Within the five U.S. districts there are three types of organizations: Defense Plant Representative Offices (DPROs) at major contractors, Residencies with smaller capabilities at some large contractors, and Defense Contract Management Area Offices (DCMAOs) that service a large number of sub-contractors.

DCMC functions are delineated in FAR Part 42.3, and are quite extensive. One of the functions we are already familiar with is that of the ACO. In addition to the ACO, DCMC provides insight into contractor purchasing systems, manufacturing surveillance, quality assurance, C/SCSC, Work Measurement practices, engineering support, and subcontracting management, just to name some. The DCMC has reorganized their internal structure to be more in-line with their customer's organizations, and have adopted the policy of being responsive to their customer's, the Program Manager's needs. Each program now has a designated Program Integrator that acts as the point of contact for the Program Manager.

The DCMC gets involved with our program to the extent recorded in the MOA between the two organizations. The primary DPRO we deal with can also issue some secondary delegations to other DCMC organizations, potentially giving us some better insight into what is happening at the various sub-contractors.

9.4   Government Contract Law

Ultimately all Federal acquisition is circumscribed by a vast body of government contract law. These laws are derived from a multitude of sources. They include statutes passed by Congress; Executive Orders; and decisions by the Federal Courts, the Comptroller General of the United States, and by Administrative Boards such as the General Services Board of Contract Appeals (GSBCA) and the Armed Services Board of Contract Appeals (ASBCA). In addition, acquisitions by Government contracting personnel are bound by official regulations which have the force and effect of law.

These are promulgated by various agencies and subagencies of the executive branch of the Government, from the Office of Federal Procurement Policy (the Federal Acquisition Regulations or FAR) down through military command regulations (Defense FARs, Air Force FARs, and Air Force Materiel Command FARs). Knowledge of all, literally hundreds, of these by someone in the acquisition process is vitally important to avoid mistakes or even illegal activity which may delay or invalidate a particular procurement.

Contracting with the Federal Government differs in many respects from ordinary commercial contracting. The Sovereign is not just another corporation. It has responsibilities which are unique, such as defending the nation. Because of these unique responsibilities, some ordinary commercial considerations and practices must be set aside when contracting with the Government. For instance, development or procurement of a weapons system must not be delayed while a contract dispute is litigated. Performance of a private contract is held up during dispute adjudication.

Although not complete, a general description of the major legal differences in the life cycle of an acquisition contract between Government and standard commercial contracting might be illustrative and valuable.

a. *Award:* Private commercial contracts may be awarded to anyone. A Government contractor must not be preselected. Each procurement action must be open and competitive with award going to that contractor best meeting predetermined criteria. In addition, selection may be limited according to a host of socioeconomic factors such as the Buy American Act, small or disadvantaged business set asides, etc. Money for the contract must be authorized and appropriated by Congress prior to execution of the contract and specific time limits for performance must be adhered to. Even the type of contract may be proscribed; cost plus percentage of cost contracts are statutorily forbidden.

b. *Interpretation:* In private commercial contracting, either party may write the contract documents and each may seek judicial interpretation of the contract. Federal contracts are written by the Government and ambiguities in the contract language are construed against the Government. The contractor must, however, continuously perform according to the Contracting Officer's instructions while clarification is sought. If the Contracting Officer is deemed incorrect in his/her interpretation, the contractor may receive additional benefits.

c. *Changes:* A private contract, once signed, may not be changed except by mutual agreement. The Contracting Officer may unilaterally change a

Government contract in most respects at any time. The contractor is never permitted to do so and is required to continue diligent performance of the contract as changed whether or not an adjustment in contract terms has been agreed upon.

     d. *Cancellation:* A private contractor who attempts to cancel or renege on an otherwise valid contract breaches that contract and the other party receives as damages all the profit which they would have made if the contract went to completion. A Government contract may be canceled (Terminated for Convenience) by the Government only, at any time prior to completion, whether or not this power is stipulated in the contract documents, and the contractor only receives profit to the date of cancellation.

     e. *Disputes:* Disputes or disagreements which develop between private parties may be litigated in local courts with immediate cessation of work during dispute resolution. In Government contracts, the decision resolving the dispute is made by the Government Contracting Officer, and it is a final decision unless appealed by the contractor either to the Washington DC sited U.S. Court of Federal Claims or to an administrative board. In either case, the contractor must complete performance while the dispute is litigated. Further appeals may be taken to higher U.S. courts but are extremely time consuming and costly.

     It should be apparent that Government contracting is highly regulated, fraught with peril to the legally uninitiated, with little discretion available to the Government contracting official except for the specifications or description of the item or service to be procured. Awareness of the body of law and regulations constraining governmental contracting is critical to the process and timely reference to Government counsel may be crucial to successful systems acquisition.

9.5   Summary

The contracting process–an iterative cycle of ongoing activities within the larger systems acquisition process–is an extremely important part of the way in which the Government's requirements are satisfied. It is the "bridge" between the Government and the defense industry across which requirements and money are exchanged for the expertise and effort required to perform the services or to build the items needed by the Government. While contracting is a very legalistic process with its authority and constraints tied to a plethora of laws, directives, regulations, and court cases, there are many knowledgeable and hard-working professionals in the contracting functions dedicated to getting the Government what it needs, when and where it is needed, at a fair and reasonable price.

References

[1] Public Law 98-369, Competition in Contracting Act (CICA)

[2] DoD Instruction 5000.2, Defense Acquisition Management Policies and Procedures

[3] AFR 70-15 (22 Feb 84), Source Selection Policy and Procedures

[4] AFSCR 80-15 (31 Dec 74), R&D Source Selection

[5] ASDP 800-7, Source Selection Guide

[6] Federal Acquisition Regulation (FAR), Parts 15, 16 and 17, 1984

[7] DoD FAR Sup, Parts 16 and 17, 1984

[8] Manual for Contract Pricing (ASPM No.1), 1986

[9] RFP Process Guide, AFSC/PK

[10] Federal Acquisition Regulation (FAR)

[11] DoD Far Supplement (DoD FAR SUPP)

[12] Defense Acquisition Regulation (DAR)

[13] AFR 110-9, Procurement Law

[14] ASPM No.1, Manual for Contract Pricing

[15] AFR 70-1, Do's and Don'ts of Industry Regulations

[16] AFR 70-1-5, DoD/NASA Incentive Contracting Guide

[17] AFSCR 70-7, AFSC Procurement Evaluation Panel

Chapter 10

SYSTEMS ENGINEERING

Capt. Brian W. Holmgren
Air Force Institute of Technology

10.1 Introduction

The discipline of Systems Engineering first came into being in the late 1950s with the advent of the Intercontinental Ballistic Missile program. The concept of the Intercontinental Ballistic Missile pushed the state of the art in a number of technical areas, resulting in the need to develop engineering specialists to concentrate on these advances. It was important that these engineering specialties worked together in a final product, and the need to balance these specialties created the concept of Systems Engineering. According to MIL-STD-499B,* Systems Engineering is:

> An interdisciplinary approach to evolve and verify an integrated and life cycle balanced set of systems product and process solutions that satisfy customer needs. Systems engineering:(a) encompasses the scientific and engineering efforts related to the development, manufacturing, verification, deployment, operations, support, and disposal of system products and processes, (b) develops needed user training equipments, procedures, and data (c) establishes and maintains configuration management of the system, (d) develops work breakdown structures and statements of work, and (e) provides information for management decision making.

As the systems engineering definition indicates, more than just engineers are involved in systems engineering. Along with engineers, we will find that logisticians, configuration and data managers, testers, manufacturing

---

*This chapter reflects the draft MIL-STD-499B, and assumes no major review changes.

personnel, cost analysts, users, and program managers are all involved in the systems engineering process. It is important that all individuals in a program office are familiar with the basic concepts of systems engineering, since systems engineering is extremely pervasive throughout the life cycle, but particularly in Engineering and Manufacturing Development, a phase that most individuals working in a program office will find themselves in at some point in time.

This chapter will examine the role of systems engineering in the life cycle process by first examining the purpose of systems engineering. The eight primary functions that all systems perform will be examined next. A model of the systems engineering process described in MIL-STD-499B will be examined, and then translated to a different model employed in the handbook that accompanies MIL-STD-499B. Some methodologies and products for controlling and employing the systems engineering process will be examined, culminating in a discussion of configuration management baselines. The article will conclude by examining the systems engineering review process in general, and then each review in greater detail.

10.2   Why Systems Engineering?

Systems engineering provides the basis for integrating the technical efforts of a multidisciplinary team to meet program cost, schedule, and performance objectives with an optimal design solution that encompasses the system and its associated manufacturing, test, and support processes (DoDI 5000.2, 6-A-1). DoDI 5000.2 Part 6 deals with engineering and manufacturing, while Part 6, Section A deals specifically with systems engineering purpose, policies, and procedures. The purpose behind systems engineering is described at the beginning of this paragraph. There are several policies that are equally succinct:

a. Systems engineering shall be applied throughout the system life cycle as a comprehensive, iterative technical management process to:

   1. Translate an operational need into a configured system meeting that need through a systematic, concurrent approach to integrated design of the system and its related manufacturing, test, and support processes;

   2. Integrate the technical inputs of the entire development community and all technical disciplines (including the concurrent engineering of manufacturing, logistics, and test) into a coordinated effort that meets established program cost, schedule, and performance objectives;

3. Ensure the compatibility of all functional and physical interfaces (internal and external) and ensure that system definition and design reflect the requirements for all system elements: hardware, software, facilities, people, and data; and

4. Characterize technical risks, develop risk abatement approaches, and reduce technical risk through early test and demonstration of system elements.

b. The primary roles of the Government and contractor program offices in the systems engineering process shall be management and execution, respectively.

c. The systems engineering process shall place equal emphasis on system capability, manufacturing processes, test processes, and support processes. (DoDI 5000.2, 6-A- 1,2)

These policies set the stage for systems engineering as a comprehensive process. Systems engineering is comprehensive in that it considers logistics, test, manufacturing, cost, schedule, and performance along with the areas one would normally consider to be engineering. The policy to "Integrate the technical inputs of the entire development community and all technical disciplines..." further reinforces the comprehensive nature of systems engineering, and also draws the user into the process. Comprehensive also refers to looking at the entire system, to include the internal and external interfaces. Being comprehensive also deals with managing the technical risk on the program. Since we are many times pushing the state of the art in technology, we many times have more risk in the technical arena than in any other portion of the program.

While the policy indicates that the Government's primary role is that of management, we in the Air Force can also assume the execution role. This assumption of roles most often occurs at the Ballistic Missile Office (BMO) at Norton AFB. BMO many times acts as the system integrator and will execute the systems engineering work. The relationship between the program offices and their prime contractors would be similar to the relationships between several prime contractors and an integrating contractor.

It is interesting to note the policy to balance system capability, manufacturing processes, test processes and support processes. Supportability was often mentioned along with cost, schedule, and performance as responsibilities of the system program director, but manufacturing and test processes have never been emphasized prior to this. You'll notice that the emphasis on manufacturing is brought out to the point that one of the life cycle phases, engineering and manufacturing development, was renamed from full scale development to help focus on the importance of manufacturing in the design process. The emphasis on test comes about partially from the need to better demonstrate our system's capabilities to help mitigate risk. Test processes go beyond development and deal with means of showing the system is operating properly and methods for troubleshooting problems.

In addition to the purpose and policies just discussed, DoDI 5000.2, Section 6-A, contains a number of procedures that are supposed to be carried out as part of the systems engineering effort. The first procedure states that "An effective systems engineering program will be implemented for each acquisition program. Recommended procedures are contained in MIL-STD-499." There is also direction to integrate the technical processes identified in MIL-STD-1388, "Logistics Support Analysis"; MIL-STD-1528, "Manufacturing Management Program"; DoD-STD-2167, "Defense System Software Development"; and MIL-H-46855, "Human Engineering Requirements for Military Systems, Equipment, and Facilities." The first procedure ends with the requirement to hold design reviews, with recommended procedures being located in MIL-STD-1521. MIL-STD-1521 will be replaced by a handbook that accompanies MIL-STD-499B. MIL-STD-499B is currently in draft, and began formal tri-service review in July 1992. Release is expected sometime in calendar year 1993.

The second procedure discusses four key systems engineering tasks. The first task is to translate operational requirements into design requirements. This task requires the program office to work with the user, or the user's representative, to establish feasible operational requirements and to identify the critical operational characteristics and constraints. Critical operational characteristics and constraints will probably end up in the acquisition program baseline (APB) for review at each milestone in the life cycle process. This task also requires that a disciplined requirements collection and translation methodology be used to convert the operational requirements into design requirements, and that each program office will establish a process that provides free and open information exchange between the design team members to balance design specifications, conduct trade-offs, and optimize the system design.

The second task is to transition technology from the technology base to program specific efforts. This task requires the program office to work closely with its key technology efforts, developing a technology transition approach that will define tasks and the resources required to accomplish the tasks. Additionally, transition criteria and an implementation methodology must be defined prior to transitioning into engineering and manufacturing development.

The third key task is to establish a technical risk management program. The task goes on to state that the technical risk management program should be a part of the overall risk management program, and that it needs to be conducted throughout the system life cycle. Provisions for either eliminating risk, or reducing it to acceptable levels, need to be included in the acquisition strategy. This task also requires that the effects of technical risk on cost and schedule, the risk reduction measures, the rationale and assumptions used in assigning risk ratings, and the alternative acquisition strategies will be explicitly assessed during the milestone reviews. This information will be contained as part of the Integrated Program Summary.

The final key task is to verify that the system design meets the operational need. This tasking further demonstrates DoD's push for more testing of new systems. This task requires the program office to establish a "comprehensive verification process....to integrate design analysis, design simulation, and demonstration and test." The tasking further states that all critical characteristics (generally those found in the acquisition program baseline - show stoppers) will be identified and verified by demonstration and test. These tests include, but are not limited to, operational effectiveness and suitability evaluations, and manufacturing process proofing tests. The final part of the tasking states that "analysis and simulation complement, not replace, demonstration and test." It concludes by saying that testing will be used to verify at least key characteristics used in the design analysis or simulation if total verification by demonstration and test is not feasible.

Several other procedures are mentioned in DoDI 5000.2 in addition to the four key tasks just discussed. The next procedure recognizes that there are a variety of different technical specialties that will have varying requirements, depending upon the program. System Program Directors are required to determine what support is necessary, are provided a list of DoD source documents for a variety of specialty areas, are required to ensure that the systems engineering process allocates requirements to the appropriate specialties, and that the systems engineering process will collectively analyze the design specifications, conduct tradeoffs, balance the total system requirements, and eventually establish the final configuration.

The next procedure discusses planning and control, requiring the program office to establish a comprehensive planning and control system for systems engineering that includes: engineering planning, technical performance measurement, configuration management, and technical data management. It requires that the System Program Director plans for major systems engineering events in the acquisition strategy. This procedure also requires the program to have the contractor submit a systems engineering management plan that addresses: (1) Management of the systems engineering process, (2) Integration of the required technical specialties, (3) Performance measures development and reporting, including intermediate performance criteria (technical performance measurement), and (4) Key engineering milestones and schedules. This procedure also requires that performance measures be developed and maintained throughout the development process to determine how well the evolving design is meeting the requirements, and those requirements that are critical to risk in particular. The procedure indicates an evolution of verification methodologies, with measures first being based on engineering judgement, and then progressing through design analysis, test data, and finally, operational data. The program must use configuration management to provide a complete audit trail on decisions and design modifications, and to track the design status to ensure valid test results. Finally, this procedure states that technical data is the formal product of the systems engineering process, and that the appropriate level of design detail must be documented throughout the development process. It discusses the fact that this data starts as validated operational requirements, moving to performance objectives and thresholds, becoming detailed design requirements, and finally ending up as specifications, drawings, process specifications, acceptance test procedures, and technical manuals.

The final procedure requires the use of a work breakdown structure to document the results of the systems engineering analysis into a structure of products and services. It goes on to state that the work breakdown structure provides the framework to relate statements of work, contract line items, configuration items, technical and management reports, and the hardware, software, and data elements of the system. It is interesting to note that even though the work breakdown structure is mentioned in systems engineering, it is a tool controlled by the financial management community, with the chair of the DoD Cost Analysis Improvement Group being the focal point. This point is made because the uses of the work breakdown structure may be viewed differently by the engineering and financial management communities. The financial management community wants the standard definitions found in MIL-STD-881 used to the maximum extent possible to gain better cost data. Engineering, on the other hand, desires to customize many of the items to a particular acquisition, and would also like to break out processes from the products to hopefully gain better insight into the processes. A current compromise in this area maintains the level 1 and 2 definitions found in the MIL-STD, allows for changes approved by OSD at level 3 and 4, and allows the program office freedom to do what they desire below level 3.

MIL-STD-499B content was developed based on the policy statements in DoDI 5000.2 Section 6-A, as well as the four key tasks from the same section, both of which are discussed above. MIL-STD-499B describes how to conduct a systems engineering program, and defines a systems engineering process model. The systems engineering process described in MIL- STD-499B is iterated over each life cycle phase, and is applied to eight primary system life cycle functions to define, design, and verify the system elements that make up the system products and processes to satisfy customer needs. You are already aware of the life cycle process from other chapters, but an explanation of the eight primary functions will be useful prior to examining the systems engineering process model.

10.3   Primary System Life Cycle Functions

By considering the development of a system in terms of eight primary system functions we should be able to better balance all of the aspects of a system in moving from requirements to disposal. The eight primary functions are: development, manufacturing, verification, deployment, operations, support, training, and disposal. The definition of each function from MIL-STD-499B will help give a deeper appreciation of what each function contributes.

(1) *Development Function:* The planning and execution of the definition, design, design implementation, integration, analyses, and control tasks, actions, and activities required to evolve the system from customer needs to product and process solutions. Development applies to new developments, product improvements, and modifications as well as any assessments needed to determine a preferred course of action for materiel solutions to identified needs, deficiencies, or problem reports.

(2) *Manufacturing Function:* The tasks, actions, and activities to be performed, and system elements required, for fabrication and assembly of engineering test models and brassboards and low-rate initial production, and full-rate production of system end items. It provides for definition of manufacturing methods and/or processes, and for the fabrication, assembly and checkout of component elements including test equipment, tooling, machinery, and manufacturing layouts.

(3) *Verification Function:* The tasks, actions, and activities to be performed, and the system elements required to evaluate progress and effectiveness of evolving system products and processes, and to measure specification compliance. Analysis (including computer analysis), demonstration, test, and inspection are verification approaches used to provide information to evaluate: risk, compliance with specification requirements, product and process capabilities, proof of concept, and warranty compliance. Included

are technology verification, manufacturing process proofing, quality assurance and acceptance, and developmental test and evaluation.

(4) *Deployment Function:* The delivery tasks, actions, and activities to be performed, and system elements required to initially transport, receive, process, assemble, install, test, checkout, train, operate and, as required, emplace, house, store, or field the system into a state of full operational capability.

(5) *Operations Function:* The tasks, actions, and activities to be performed, and the system elements required to satisfy defined operational objectives and tasks in the peacetime and wartime environments planned or expected.

(6) *Support Function:* The tasks, actions, and activities to be performed, and the system elements required to provide operations, maintenance, logistics (including training) and materiel management support. It provides for the definition of tasks, equipment, skills, personnel, facilities, material, services, supplies, and procedures required to ensure the proper supply, storage, and maintenance of a system end item.

(7) *Training Function:* The tasks, actions, and activities to be performed, and the system elements required to achieve and maintain the knowledge and skill levels necessary to efficiently and effectively perform operations and support functions.

(8) *Disposal Function:* The tasks, actions, and activities to be performed, and system elements required, to ensure that disposal of decommissioned and destroyed or irreparable system end items comply with applicable classified and environmental regulations and directives. Additionally, disposal addresses the short and long term degradation to the environment and health hazards to humans and animals. The disposal function also includes recycling, material recovery, salvage for reutilization, and disposal of by-products from development and production.

Even though we have these eight neatly packaged functions, we must also worry about things like availability, reliability and maintainability, the integrated logistic support elements, and producibility. Most of these considerations are addressed directly in the format for the operational requirements document found in DoDI 5000.2-M, and are implicit to the definition for their corresponding functions in the preceding paragraphs. Now we will examine the systems engineering process model.

10.4   Systems Engineering Process Model

According to MIL-STD-499B, the systems engineering process is :

> A comprehensive, iterative problem solving process that is used to: (a) transform validated customer needs and requirements into a life cycle balanced solution set of system product and process designs, (b) generate information for decision makers, and (c) provide information for the next acquisition phase. The problem and success criteria are defined through requirements analysis, functional analysis/allocation, and systems analysis and control. Alternative solutions, evaluation of those alternatives, selection of the best life cycle balanced solution, and the description of the solution through the design package are accomplished through synthesis and systems analysis and control.

The systems engineering process is applied in each phase of the acquisition life cycle, although the emphasis will change from phase to phase. Before we examine how the process is employed in each phase, let's examine the parts of the model in greater detail. In addition to the model steps mentioned in the definition, we also have input to, and output from the process. As we look at each part of the model, we must keep in mind that the steps in the process are performed iteratively with each other. That is, we don't perform requirements analysis and then move to functional analysis/allocation and never look back. In much the same manner, we don't perform requirements analysis without doing some thinking ahead also. There will be a number of potential trade-offs throughout the process that can only be adequately addressed by iterating through the model. A schematic representation of the systems engineering process is shown in Fig. 10.1.

INPUT

A number of inputs go into the systems engineering process, and they will vary considerably from phase to phase. Inputs start initially with the mission need statement, which will state the user's need in terms of broad operational objectives, as well as known constraints and desired operating capability in specified environments. This initial input becomes more refined as the system moves further into the life cycle. Other inputs come from the technology base, directions and limitations from the milestone decision authority, refined requirements from the operational requirements document, requirements applied through standards and specifications, and output from the previous phase.

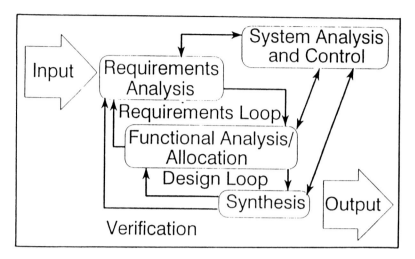

Fig. 10.1: Systems engineering process.

REQUIREMENTS ANALYSIS

Requirements analysis serves as the bridge between user requirements and system specific requirements from which solutions can be generated. Requirements analysis is performed by analyzing user needs, objectives, and requirements in the context of missions, utilization environments, and identified system characteristics. Requirements analysis results in the determination of functional and performance requirements for each of the eight primary functions of the system. A portion of this analysis is concerned with reviewing and updating any prior analyses that have been accomplished to help support defining the system. Requirements analysis is conducted iteratively with functional analysis/allocation and synthesis to: (a) develop requirements that depend on additional system definition, and (b) verify that product and process solutions can satisfy customer requirements. Desired outcomes of the requirements analysis process include: (1) refined customer objectives and requirements, (2) defined initial performance objectives that are further refined into requirements, (3) identified constraints in terms of environments (utilization and human) and interoperability, as well as others, and (4) functional and performance requirements based on user supplied measures of effectiveness.

FUNCTIONAL ANALYSIS/ALLOCATION

Functional analysis/allocation is conducted to define and integrate a functional architecture for which system products and processes can be designed. Functional analysis/allocation is usually performed in a top-down fashion, first identifying major functions that a system must accomplish, and then decomposing those functions to lower level functions. Each succeeding function identified in the decomposition process is further decomposed until a level of detail is reached that allows for the successful accomplishment of the synthesis of solutions. The eight primary functions discussed earlier provide the basic framework to start functional analysis/allocation from, but must be decomposed to lower levels of detail to gain the information necessary to conduct synthesis. Tools like functional flow block diagrams and time line analysis are commonly employed as part of this process. The "Systems Engineering Management Guide" from the Defense Systems Management College contains an excellent discussion of these tools, although their process model differs from the model being presented here, and some of the material does not reflect the current acquisition environment.

As part of this process derived requirements will be developed. Derived requirements are requirements necessary to ensure that stated requirements in the system specification can be met. Derived requirements can be in terms of performance, like a specification of accuracy in an inertial navigation element's calculations to support the system level circular error probable (CEP) calculation, or in physical terms, such as dimensions of avionics boxes as space is allocated in the system for all of the necessary components. Derived requirements will often show up in the configuration item specifications, but not in the system level specification since they support defined system level requirements.

As alluded to in the last paragraph, once functions have been decomposed to lower levels, requirements are allocated to proposed potential configuration items. Many of these potential configuration items come from the work breakdown structure initially found in MIL-STD-881. Other potential configuration items may come from previous systems engineering work. This requirements allocation is crucial to ensuring that the system will eventually work as envisioned, and it is a step that is not always as critically examined as it should be. It is absolutely crucial to establish that all of the system level requirements, and derived requirements, have been either allocated to one or more configuration items, or have been knowingly held at the system level. In addition, it is also crucial to determine if the requirements are written so they are testable, preferably using the measure of effectiveness that the user has identified. Once this part of the process is complete, remembering that it is iterative, synthesis will be accomplished on the identified configuration items.

SYNTHESIS

Synthesis is the part of the process that defines solutions using a bottoms-up methodology. The purpose of synthesis is to define and design system product and process solutions in terms of design requirements that satisfy the functional architecture and define and integrate the system as a physical architecture. Synthesis is conducted iteratively with functional analysis/allocation to define a complete set of functional and performance requirements necessary for the level of the design output required (phase dependent), and with requirements analysis to verify that solution outputs can satisfy customer input requirements.

SYSTEMS ANALYSIS AND CONTROL

Systems Analysis and Control is performed to provide the progress measurement, assessment, and decision mechanisms required to evaluate design capabilities and document the design and decision data. Systems analysis includes trade studies, effectiveness analyses and assessments, and design analyses to determine progress in satisfying technical requirements and program objectives, and to provide a rigorous quantitative basis for performance, functional, and design requirements. Systems analysis is used to support requirements analysis, functional analysis/allocation, synthesis, trade studies, risk management, and technical performance measurement. Control is obtained through risk management, configuration management, data management, and progress based measurement efforts consisting of the systems engineering master schedule, technical performance measurement, and technical reviews. The systems analysis and control effort ensure that: (1) decisions on solution alternatives are made only after evaluating their impact on system effectiveness, life cycle resources, risk, and user requirements, (2) engineering decisions, and system unique specification requirements, are traceable to systems engineering activities, (3) traceability from process inputs to process outputs is maintained, including changes in requirements and results of decisions being documented, (4) schedules for the development and delivery of products and processes are mutually supportive, (5) technical disciplines and disciplinary efforts are integrated into the systems engineering process execution, (6) impacts of user requirements on resulting functional and performance requirements are examined for validity, consistency, desirability, and attainability with respect to technology availability, physical and human resources, human performance capabilities, life cycle costs, schedule, risk, and other identified constraints. The results of this examination will either confirm existing requirements, or require determining more appropriate system requirements that can be iterated with the user, and (7) product and process design requirements are directly traceable to the functional and performance requirements they were designed to fulfill, and vice versa.

OUTPUTS

Outputs from the systems engineering process are phase dependent, and are normally stored in some form of decision data base so they can be accessed as necessary to have a complete audit trail of the process. The output from the systems engineering process encompasses the data necessary for system life cycle management, including information on the functional and physical architecture, specifications, configuration baselines, technical inputs to acquisition documents, technical data required to support decisions, and technical data necessary to operate and maintain the system.

10.5   Alternate Systems Engineering Model

A systems engineering/configuration management process action team met through most of 1991, and all of 1992, to provide the framework for systems engineering and configuration management as it would be utilized by Air Force Materiel Command. Membership included individuals from both Air Force Logistics Command and Air Force Systems Command. Some of the members of the process action team were concurrently involved in the preparation of MIL-STD-499B, so the processes have evolved simultaneously, and somewhat in conjunction with each other.

One of the desired outcomes of the process action team was the development of a handbook that provided a look at how to apply systems engineering throughout the entire life cycle process. After months of iterating and refining, the process action team finalized a process model that covered 180+ blocks and extended over thirteen pages of flow diagrams. Detailed explanations of each block in the process were developed, and all of the products were placed into a draft handbook to accompany MIL-STD-499B.

According to this representation, the entire life cycle flow consists of a basic "Systems Engineering Engine" that is applied in each life cycle phase, along with various considerations that are phase specific. The emphasis of the "Systems Engineering Engine" varies with each life cycle phase, although the engine itself remains unchanged. The "engine" consists of a stack of three blocks as shown in Fig. 10.2. You will notice that there are some differences between this model and the one represented in Fig. 10.1, although they both represent the same types of analyses. In order to connect the handbook to MIL-STD-499B, it is necessary to construct a mapping between the two model representations.

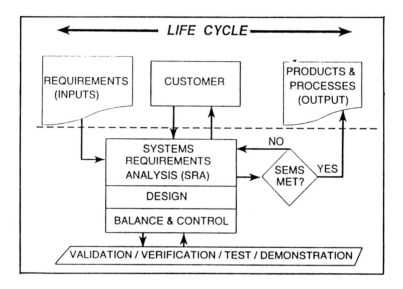

Fig. 10.2: SE/CM top level process.

Several blocks map directly from the SE/CM Top Level Process to the MIL-STD-499B model, as is the case for Inputs and Outputs. The Systems Requirements Analysis (SRA) process is a composite of the requirements analysis and functional analysis/allocation tasks found in MIL-STD-499B. The Design Block maps directly to the MIL-STD-499B Synthesis task. The Balance and Control Block and SEMS Met? Block correspond to the system analysis and control tasks of MIL-STD-499B. It is important to note that the inputs and outputs flow into the "engine", or stack of three blocks, in the SE/CM model rather than just into the SRA block as it might appear. This is consistent with the way the MIL-STD-499B process is depicted.

As you can see, the mapping between the two models is fairly straightforward, and greater detail is available in the handbook. Now that the models have been examined, it is important to know the emphasis of the systems engineering process in each phase of the life cycle process.

10.6   Life Cycle Application of the Systems Engineering Process

Using the model found in the handbook to MIL-STD-499B, it was stated previously that the emphasis of the "systems engineering engine" varied with the phase of the life cycle the program was in. The process defined in the MIL-STD-499B handbook extends from pre- concept exploration and definition through operations and support. Understanding the emphasis of each phase will help you better know how to interact with the systems engineering process and what to reasonably expect to see in terms of results.

PRE-CONCEPT EXPLORATION AND DEFINITION PHASE

The major emphasis in the Pre-Concept Exploration and Definition phase is on aiding the using commands with threat analysis, mission needs analysis support, identifying viable technological solutions, formulation of initial performance parameters, risk analysis and mitigation, and identification of technology strategies in support of the user's future perceived needs. This is the period of time in the life cycle in which the mission need statement (MNS) is formulated. One of the major goals of this phase is to identify future needs that the user may have and to conduct research programs that will help foster new technologies to reduce risk when the technologies are needed.

CONCEPT EXPLORATION AND DEFINITION

The systems engineering process must now respond to the mission need statement that the user has generated. The main emphasis in Concept Exploration and Definition is to develop alternative concepts designed to meet the need, and then to evaluate those alternatives to determine which concept seems to be most reasonable to take into Demonstration and Validation. The systems engineering process will be supporting the user in defining the operational requirements document (ORD) and in performing a cost and operational effectiveness analysis (COEA). Each alternative concept will have to be analyzed in terms of the eight primary functions and the technologies available to meet the requirements of each alternative. It will be necessary to identify technologies that will have to be invested in to help mitigate the risk. Additionally, the systems engineering process will need to help define the requirements for prototypes in Demonstration and Validation, as well as the criteria to assess risk mitigation, some of which will become exit criteria. An Alternative System Review (ASR) is normally conducted during this phase.

DEMONSTRATION AND VALIDATION

The Demonstration and Validation phase is when the systems engineering process has normally provided the data necessary to narrow the alternative concepts to a single best concept, and contractors are developing alternative designs to best meet the requirements for the selected concept. The main goals of this phase are to mitigate risk through technology demonstration in the prototypes, to better define and refine the requirements in the Operational Requirements Document, and to evolve the system analysis to a sufficient point that the initial allocation of requirements to proposed configuration items has been completed in enough detail to ensure technologies are available, or that they are being matured to reduce risk. Ideally the functional view, or system level requirements, will be completed during this phase so the winning contractor will be able to refine their functional analysis/allocation in Engineering and Manufacturing Development and move on to the next iteration of synthesis tasks. Successful requirements identification over the eight functions, and then along the missions that the system is to accomplish within those eight primary functions, is crucial to the eventual technical success of the program in the most economical manner. Unfortunately, with the complexity of today's weapons systems, it is also one of the most difficult tasks to accomplish thoroughly. There will need to be intense collaboration between the user, program office, and air logistics center to best define the total system requirements. There are two major reviews in this phase, the System Requirements Review (SRR) and the System Functional Review (SFR).

ENGINEERING AND MANUFACTURING DEVELOPMENT

Engineering and Manufacturing Development is when configuration item requirement allocation is finalized and design solutions are translated into hardware and software that meet the user's need. While prototypes that primarily focus on the prime piece of equipment are designed as rapidly, and as inexpensively, as possible to demonstrate critical technologies and mitigate risk, full systems must be thoroughly developed during engineering and manufacturing development. All of the things that support the prime piece of equipment are developed, as well as the full up prime piece of equipment.

Testing becomes particularly crucial during engineering and manufacturing development as we try and determine if the newly designed system will be able to meet user needs. Testing begins first on individual configuration items, and then it builds to where a full up, hand built, prime piece of equipment is ready for development testing. At the same time all of that is happening, the contractor is also trying to develop support equipment, some of which could be as complex as the prime piece of equipment, in order to support the initial operational test and evaluation that first looks at a production representative system to see if it meets the user needs.

Did we say production representative system in that last paragraph? Since we did, we must realize that the contractor has to also be developing his manufacturing capacity in parallel, or concurrently, with the prime piece of equipment and the support equipment. The manufacturing considerations extend back to concept exploration and definition to ensure that producibility impacts the design. Manufacturing is validated in Low Rate Initial Production (LRIP) which also supplies the production representative systems for initial operational test and evaluation. In some newer programs, notably the B-2, production tooling was developed concurrently with the aircraft design, and there were no hand-built products, only products that came directly from the production tooling.

While only three processes that occur concurrently are highlighted here, there are many additional processes also occurring during engineering and manufacturing development. Just think about all of the complexities of interfacing the activities we've already discussed, and then developing interfaces with other systems our system must work with. This brief glimpse of engineering and manufacturing development will give you a better idea of the complexity involved in systems engineering during this particular phase, and why multidisciplinary teams that consider all functions concurrently are so important.

PRODUCTION AND DEPLOYMENT/OPERATIONS AND SUPPORT

Although Production and Deployment, and Operations and Support, are actually different phases of the life cycle process, they have been lumped together for systems engineering purposes because the same types of things happen in both phases. The primary concern in systems engineering is handling change. Change may occur as a result of operational problems or evolution of mission. In order to handle these changes properly, it is important to use the same disciplined set of steps used in previous phases to best determine how to handle the modification regardless of what caused the need for it. The necessity for multidisciplinary teams looking at all eight primary functions is not diminished in this phase. Documentation to control the process must be kept current as well, a problem that is compounded by the user servicing equipment and making undocumented changes to keep equipment operational.

10.7   Employing and Controlling the Systems Engineering Process

There are several tools used to employ and control the systems engineering process, things like systems engineering management plans, systems engineering master schedules, systems engineering detailed schedules, technical performance measurements, configuration management baselines, and a series of technical reviews.

SYSTEMS ENGINEERING MANAGEMENT PLAN (SEMP)

The Systems Engineering Management Plan (SEMP) defines the conduct and management of the fully integrated engineering effort necessary to satisfy the general and detailed requirements of MIL-STD-499B. Both the Government and the contractor will develop systems engineering management plans to show how each will handle the portions of the systems engineering process they are responsible for.

Government systems engineering management plans should include: (1) a general life cycle roadmap of key systems engineering activities, the responsible Government office for each activity, as well as tailoring approaches for MIL-STD-499B and any other specifications and standards, (2) Government multidisciplinary team structures and responsibilities, (3) plans and criteria for transitioning critical product and process technologies, and (4) identification of key trade studies, the scope and depth of systems effectiveness assessments, the current measures of effectiveness hierarchy, technical risk management plans, critical technical parameters, and tracking requirements for those parameters.

The contractor systems engineering management plan addresses how the contractor will manage three essential characteristics of concurrent engineering: (1) the simultaneous development of system products and life cycle processes to meet user needs, (2) the utilization of multidisciplinary teams, and (3) a systems engineering methodology. The systems engineering management plan not only contains what the contractor plans to do to accomplish systems engineering, but also how the efforts will be accomplished, who will accomplish them, how they will be controlled, and how technology will be transferred from the technology base to system products and processes. MIL-STD-499B recommends giving some consideration to making the SEMP a contractual document. In addition to the systems engineering management plan, the contractor will normally be required to prepare a systems engineering master schedule.

SYSTEMS ENGINEERING MASTER SCHEDULE (SEMS)

The systems engineering master schedule is developed by the contractor to identify significant accomplishments that must be achieved by established contract events. The contractor must include, at minimum, the events, accomplishments, and associated success criteria identified by the Government. All events and accomplishments are tied to work breakdown structure elements. Events are identified either in the format of entry and exit events (e.g., Initiate PDR and Complete PDR) or by using entry and exit criteria for each event. Listed accomplishments are supposed to be event related and not time coincidental or driven. Accomplishments should have one or more of the following characteristics: (1) defines a desired result at a specified event that indicates design maturity or progress directly related to each product and process, (2) defines completion of a discrete step in the progress of the planned development, or (3) describes the functional activity directly related to the product.

Systems engineering master schedule (SEMS) criteria should be measurable, like: (1) test plan complete, (2) safety significant item list finalized, (3) supportability requirements implemented in design, or (4) achievement to date of a technical parameter within technical performance measurement tolerance band, and current estimate satisfying requirement threshold, instead of being unmeasurable, like "test plan 85% complete." Finally, the Government should identify critical technical performance measurements to be used as accomplishment criteria for identified milestones. These parameters should be risk based and be tied to either the acquisition program baseline (APB) or exit criteria. In addition to a systems engineering master schedule, the contractor has to show the program office how they will accomplish their efforts through a systems engineering detailed schedule.

SYSTEMS ENGINEERING DETAILED SCHEDULE (SEDS)

The systems engineering detailed schedule (SEDS) is a time based schedule that shows how the contractor is planning their work efforts to support the events and tasks identified in the systems engineering master schedule (SEMS). The systems engineering detailed schedule outlines the tasks and calendar dates necessary to show when each significant accomplishment will be achieved. The contractor is required to maintain the relationship between the SEMS and the SEDS.

TECHNICAL PERFORMANCE MEASUREMENT (TPM)

Technical performance measurement is the continuing verification of the degree of anticipated and actual achievement of technical parameters. Technical performance measurement is used to identify and flag the importance of a design deficiency that might jeopardize meeting a system level requirement that has been determined to be critical. Measured values that fall outside an established tolerance band require corrective action. Figure 10.3 illustrates a technical performance measurement.

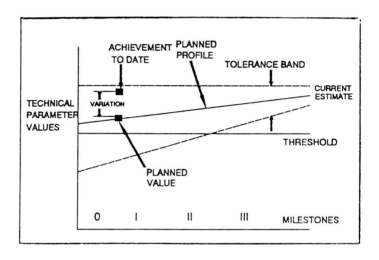

Fig. 10.3: Technical performance measure profile.

Some of the relevant terms dealing with technical performance measurement are:

1. Achievement to Date: Measured or estimated progress plotted and compared with planned progress by designated milestone date.

2. Current Estimate: Expected value of a technical parameter at contract finish.

3. Planned Value: Predicted value of parameter at given point in time.

4. Planned Profile: Time phased projected planned values.

5. Tolerance Band: Management alert limits representing projected level of estimating error.

6. Threshold: Limiting acceptable value, usually contractual.

7. Variation: Difference between the planned value of the technical parameter and the achievement-to-date derived from analysis, test, or demonstration.

CONFIGURATION MANAGEMENT BASELINES

Another measure of control utilized in systems engineering is configuration management baselines. Configuration management baselines allow the Government to take responsibility for the system on an incremental basis. There are other tasks in configuration management that will be covered in a separate paper, but configuration management baselines need to be discussed here because of their importance in first taking control of the system level requirements, and then eventually the drawings.

The first baseline is the Functional Baseline. It is the initially approved documentation describing a system's, or Configuration item's, functional, performance, interoperability, and interface requirements, and the verification required to demonstrate the achievement of those specified requirements. This baseline is normally placed under government control during Demonstration and Validation.

The second configuration management baseline is the Allocated Baseline. The allocated baseline is the approved documentation describing a configuration item's functional, performance, interoperability, and interface requirements that are allocated from the system or higher level configuration item; interfacing requirements with interfacing configuration items; design constraints; derived requirements (functional and performance); and verification requirements and methods to demonstrate the achievement of those requirements and constraints. The allocated baseline is typically place under government control during engineering and manufacturing development. There is an allocated baseline for each configuration item.

The final baseline is the Product Baseline. The product baseline consists of the approved documentation describing all of the necessary functional, performance, and physical requirements of the configuration item; the functional and physical requirements designated for production acceptance testing; and tests necessary for deployment, support, training, and disposal of the configuration item. This baseline normally includes product, process and material specifications, engineering drawings, and other related data. In addition to the documentation, the product baseline of a configuration item may consist of the actual equipment and software. This baseline is typically placed under control after completion of the physical configuration audit (PCA), which usually occurs either late in engineering or manufacturing development, or very early in production and deployment. There is a product baseline for each configuration item.

TECHNICAL REVIEWS

The last method of controlling the systems engineering process is through a series of technical reviews. Technical reviews should not be viewed as a single point in time, but as a culmination of a significant portion of the technical effort. Reviews may be conducted on an incremental basis during the technical effort, or may be held at the end of the effort; something that is determined by the program office. Reviews are currently discussed, and governed, by MIL-STD-1521 until such time when the MIL-STD-499B Handbook on Technical Reviews is published. The reviews that will be present under the new MIL-STD-499B are discussed in this chapter. The systems engineering master schedule (SEMS) should establish entry and exit criteria for each review. Some of the entry and exit criteria are discussed in this paper, and Appendix C of draft MIL-STD-499B contains all of the criteria. Functional and physical configuration audits are not discussed in this paper, but will be left for the lesson on configuration management.

In order for technical reviews to be effective, they must be attended by the multidisciplinary team working the particular part of the program, and the attendance should be kept down to the minimum number of necessary personnel. It is important to review the statement of work, the operational requirements document, the system specification, as well as any additional documentation that is presented in connection with the review being attended. By having all team members be familiar with the documentation, they should understand what the contractor is, and isn't supposed to be doing. This can save a tremendous amount of time and allow the team to focus on the real issues. Personnel need to be educated about what to expect from the contractor at the review so we aren't asking for details that are premature. A team meeting that focuses on the issues team members have prior to the actual review is invaluable in helping to focus team members on the issues. This team meeting also ensures that we have a chance to discover the issues and reach some concurrence among team members prior to the review instead of either having to work out differences at the contractor's facility or missing issues altogether.

There are several crucial things to remember when the review is in progress. First of all, remember to put up a disclaimer statement to make sure everyone is aware that no direction will come from the review personnel, only from the contracting officer. This disclaimer should help minimize the potential for constructive change. Government caucuses should be held when necessary, and especially when we have disagreement on the Government side of things. As problems are noted, they must be documented in action items. Action items must then be closed as soon as possible after the review, otherwise they may never get closed. Permanent action items may go away if the draft policy in AF Sup 1/DoDI 5000.2 gets signed. According to the draft supplement, design reviews are an excellent chance to practice risk management, and it prohibits conducting the next review until all action items from the last review are closed.

10.8 System Engineering Reviews

Now that we have seen what we need to do to get ready for reviews, and what we must watch out for at reviews, we will look at the reviews at greater detail and in chronological order.

ALTERNATE SYSTEM REVIEW (ASR)

The Alternate System Review (ASR) is a new review under MIL-STD-499B, and replaces the System Requirements review that used to occur during Concept Exploration and Definition. The major purpose of this review is to determine if a concept meets the identified need, and then to assess the concept in enough detail to make a decision about which concept to select. This review covers much of the same information that the System Requirements Review used to in this phase, but there should be a greater emphasis on assessing the future activities necessary to reduce the risk of the preferred concept to a point where commitment to a system specification is feasible. Part of this review may be having functional reviews in areas like support, training, manufacturing, and any other desired areas, to determine the functional issues with a particular concept and ensure they are considered in system planning. There may be multiple ASRs if more than one contractor is conducting concept studies.

Entry accomplishments include: (a) concept studies complete, (b) concept system architectures complete, (c) analytic assessments that the concepts meet the need, (d) estimates of cost, schedule, and performance thresholds and objectives complete, (e) product and process technology verification requirements for future developments complete, and (f) product and process risks identified and risk management approach complete.

Exit accomplishments include: (a) draft specification tree and program work breakdown structure for Demonstration and Validation complete, and (b) technical exit criteria for Demonstration and Validation defined. Information from this review will feed into the Cost and Operational Effectiveness Analysis (COEA) which will recommend the preferred concept.

SYSTEM REQUIREMENTS REVIEW (SRR)

The System Requirements Review (SRR) is now conducted at the beginning of Demonstration and Validation. Alternatively, this review could be conducted in Production and Deployment, or in Operations and Support for modifications, upgrades, and product and process improvements. The purpose of this review is to ensure that all parties involved understand the system level requirements on a mutual basis.

Entry accomplishments include: (a) draft system specification complete, (b) draft system architecture complete, (c) nonviable candidate technologies identified.

Exit accomplishments include: (a) contractor understanding of customer requirements, and (b) draft functional baseline complete.

SYSTEM FUNCTIONAL REVIEW (SFR)

The System Functional Review (SFR) is new under MIL-STD-499B, and is actually a refocus of the System Design Review found in MIL-STD-1521B. The major purpose of the review is to establish, and verify, an appropriate set of functional and performance requirements for the system. The system functional review is normally held late in Demonstration and Validation, but may be held at the beginning of Engineering and Manufacturing Development for programs not going through Demonstration and Validation, or in Production and Deployment or Operations and Support for modifications, upgrades, and product and process improvements.

Entry accomplishments include: (a) all scheduled subsystem, functional, and interim system reviews complete with all unresolved issues documented, (b) definitization of functional and performance requirements for the eight primary functions, (c) system level trade studies completed, (d) configuration item identification trade studies complete, (e) system architecture, specification tree, and draft configuration item architecture complete to an agreed to measurable level, (f) draft configuration item development specifications complete to an agreed to number of levels below the system specification, (g) effectiveness assessments (to support specifications) complete, and (h) establishment of configuration management program complete.

Exit accomplishments include: (a) system specification complete, (b) functional baseline for products and processes complete, (c) draft contractor allocated baseline complete, (d) draft interface control documents complete, (e) proposed final program work breakdown structures complete, (f) pre-planned product improvement and evolutionary acquisition strategies complete, and (g) risk handling approaches for Engineering and Manufacturing Development complete. In addition, another criteria might be establishing technical exit criteria for Engineering and Manufacturing Development.

PRELIMINARY DESIGN REVIEW (PDR)

The Preliminary Design Review (PDR) remains the same as it always has. The primary purpose for the PDR is to assess the progress of the evolving design, and to ensure that all of the allocated requirements for each configuration item are well understood. A Preliminary Design Review is conducted for each configuration item, or aggregation of similar configuration items, prior to having a system level PDR. This series of reviews is normally conducted early in Engineering and Manufacturing Development, but may occur in Demonstration and Validation for major prototyping activities, or in Production and Deployment or Operations and Support for modifications, upgrades, and product and process improvements.

Entry accomplishments include: (a) all scheduled subsystem, functional, and interim system reviews complete with all unresolved issues documented, (b) system architecture update complete, (c) preliminary configuration item architecture complete, (d) preliminary designs complete, (e) all functional and physical interface requirements established, (f) design implementation trade studies complete, (g) make/buy decisions finalized, and (h) compatibility between all configuration items established.

Exit accomplishments include: (a) verification that the functional and performance requirements of each configuration item satisfy the functional baseline, (b) updated pre- planned product improvement or evolutionary acquisition strategies completed, (c) agreement that requirements for all configuration item requirements are satisfied by the design approach, (d) all draft functional and physical interface control documents completed, and (e) preliminary allocated baseline completed for each configuration item.

CRITICAL DESIGN REVIEW (CDR)

The Critical Design Review (CDR) is also unchanged under MIL-STD-499B. The primary purpose for the CDR is to assess the progress of the evolving design, and to ensure that the detailed design satisfies all the allocated requirements for each configuration item, and that the contractor is ready to start building the system for developmental testing. A Critical Design Review is conducted for each configuration item, or aggregation of similar configuration items, prior to having a system level CDR. This series of reviews is normally conducted early in Engineering and Manufacturing Development, but may occur in Demonstration and Validation for major prototyping activities, or in Production and Deployment or Operations and Support for modifications, upgrades, and product and process improvements.

In circumstances like the Clear Accountability in Design initiative, where the Government has elected to not bring the allocated baseline under configuration control at this review, or earlier, an assessment of the flowdown of requirements from the functional baseline to the lowest level configuration item for each item in the specification tree should be accomplished at this review. Any changes in the contractor's draft allocated configuration documentation since the Preliminary Design review are reviewed by the Government, and their impact to the functional baseline is assessed and validated.

Entry accomplishments include: (a) all scheduled subsystem, functional, and interim system reviews complete with all unresolved issues documented, (b) system architecture update complete, (c) update of system specification complete, (d) update of functional baseline complete, (e) all configuration item architectures complete, (f) all interface control documents complete, (g) detailed designs complete, (h) all configuration item draft product specifications complete, and (i) all configuration item draft product baseline complete.

Exit accomplishments include: (a) verification that the functional and performance requirements of each configuration item satisfy the functional baseline, (b) update of allocated baseline completed (c) assessments confirm that configuration item and system requirements are satisfied by the design, and (d) qualification test articles are ready for fabrication/coding.

SYSTEM VERIFICATION REVIEW (SVR)

The System Verification Review is also a new review under MIL-STD-499B. The major purpose of this review is to verify that the system is ready for production. This review is normally conducted in late Engineering and Manufacturing Development, but may occur in Production and Deployment or Operations and Support for modifications, upgrades, and product and process improvements. The SVR is conducted after functional configuration audits have been performed on all configuration items.

Entry accomplishments include: (a) all system specification Part 4 verification tasks against Part 3 requirements completed, (b) functional configuration audit for each configuration item, (c) verification that each item tested conforms to its design, (d) update of all architectures and baselines, established and initiated, complete, (e) verification that the availability, capability, and capacity of manufacturing elements is complete, and (f) product and process designs are stable.

Exit accomplishments include: (a) verification of the functional characteristics of all products and processes completed, (b) all allocated baselines established, (c) production work breakdown structure define, (d) system ready to be produced.

10.9   Summary

Systems Engineering is the process that translates the user's requirements into specifications that a contractor can use to design a system. An understanding of how the requirements process interacts with, and evolves through, the systems engineering process is crucial in getting the system the user needs. Systems engineering is performed concurrently by integrated teams when it is employed properly. Integrated teams provide the system perspective necessary to properly balance all aspects of the system to provide a capability that meets the user's need and is obtained at the most reasonable life cycle cost.

This chapter provides only a brief overview of all that goes into the systems engineering process. More information can be found in DoDI 5000.2, DoD 5000.2-M, MIL-STD-499B, MIL-STD-1521, MIL-STD-973, and the new handbook for MIL-STD-499B.

References

[1] DoD Instruction 5000.2, "Defense Acquisition Management Policies and Procedures," Office of the Secretary of Defense, Under Secretary of Defense for Acquisition, 23 February 1991.

[2] MIL-STD-499B, "Systems Engineering," 24 April 1992, Draft.

[3] MIL-HDBK-499-3, "Systems Engineering/Configuration Management Life Cycle Application (Part 3 of 5 parts)," 16 December 1992, Draft.

[4] Air Force Supplement 1/DoDI 5000.2, "Defense Acquisition Management Policies and Procedures," February 1993.

Chapter 11

CONFIGURATION MANAGEMENT

11.1   Configuration Management Concepts

The previous chapter dealt with the systems engineering process and the reviews that accompany the process. Configuration management goes hand in hand with those reviews in the systems engineering process in controlling the requirements, design, and eventually drawings and physical status of items in the field.

The "configuration" of a system/item is its set of descriptive and governing characteristics that can be expressed in both functional and physical terms. They are set forth in technical documentation (such as specifications, drawings, and code listings) and achieved in the delivered product. Functional terms deal with the performance the item should achieve (how high, how far, how fast, how reliable, how supportable, how survivable). Physical terms define what the item should look like (e.g., dimensions, surface finish, color) and consist of (e.g., materials, exact part numbers) when it is built. In practice, an item under development is primarily described by, and controlled to, its functional characteristics. Once it enters production and operation, it is primarily described by, and controlled to, its physical characteristics; however, most programs continue to maintain control of the functional characteristics of the item throughout its operational life, too.

Configuration Management (CM) is the process that: (1) identifies the functional and physical characteristics of an item and acquires the documentation needed to define those characteristics during the item's life cycle, (2) controls changes to those characteristics, (3) audits the achievement or non-achievement of the required characteristics, and (4) provides an information system that accounts for the status of all approved configurations and their changes. There are the four functions that CM performs in support of an acquisition program; DoDI 5000.2, Part 9, calls them Configuration Identification, Configuration Control, Configuration Audits, and Configuration Status Accounting. When performed correctly, CM provides the following benefits to the Program Manager:

- A process for establishing a "TECHNICAL contract" for the items to be delivered under the contract for the program. (Baselines are established as the program evolves to contractually define the functional and physical requirements for the item and the related verifications to be used to assure that the requirements have been achieved.)

- A disciplined structure for controlling the submittal and approval of changes to the baselined documentation (the technical contract), along with procedures for assessing the impact of changes and for monitoring their implementation.

- An information system providing precise identification of the current configuration of all program end items, including traceability back to previously approved technical baselines and system requirements. This traceability is ESSENTIAL to successful logistics support of the operational items.

Government contracts for major acquisitions require a formal CM program, with specific requirements being levied on the contractor. The contractor is normally tasked to generate the documentation used to define the configuration; to generate the change proposals identifying potential changes to the baselined documentation; and to establish and maintain a comprehensive information system which tracks the documentation, the changes to the documentation, and the configuration of the items being delivered. At the same time, the program office must also perform many related configuration management activities, including the review of the draft documentation; the review of the change proposals; the accomplishment of the audits to verify successful achievement of the functional and physical characteristics; the establishment and maintenance of a complementary information system which tracks changes under review in the program office and the actual configuration of delivered units in the field, and the use of the information system to manage the items and documentation. The procedures for each are very similar and thereby encourage continuous communication and coordination between the program office and contractor CM offices.

Within the program office itself, the CM office has regular dealings with the business management functions in the Program Control and Procurement/ Contracts offices, with the support function in the Logistics office, and with the technical functions in the Engineering offices. CM and the CM office provide a crossfeed of information among various functional activities in the program office. The effective PM uses CM as a means of achieving required system performance (through the definition of the required performance and design), of controlling program costs (through the careful control of changes), and of maintaining schedule. When CM is ignored or poorly implemented, the PM loses his ability to define and control the technical aspects of the program; to properly manage his contractors; and to effectively deal with his user community. More importantly, poor CM usually will lead to significant problems with the logistics support of the system.

CM can be applied during all phases of the system acquisition life cycle. Although not required during the Conceptual phase (exploratory and advanced development), the generation of draft system-level specifications as a part of the Systems Engineering process in that phase is actually the start of the configuration identification process; depending on the way the program is structured, either the contractor or the government may be responsible for these initial CM activities. During the demonstration, development, and production phases, the configuration identification process is completed, the configuration audits are accomplished, and the configuration control and configuration status accounting processes are begun; much of the responsibility related to these activities rests with the contractor. During the operational phase, CM consists primarily of the control and status accounting processes, although updates and revisions to the documentation are required; however, the responsibility rests with the government management activity.

11.2   Configuration Items

CM activities during the development of the design are focused on system elements called Configuration Items (CIs). These are logical groupings of hardware and/or software elements having similar performance and design requirements which satisfy a recognizable (and usually a significant) end-use function. A list of candidate CIs is one of the required outputs of the Systems Engineering process. During development, each acquisition program has at least two CIs: the primary hardware element and the primary software element of the system being acquired. As the size and complexity of the system and its component elements increases, it is more likely that these major system elements will be subdivided into additional CIs. The designation of CIs is a management decision made by the program office based on judgement, experience and various engineering and logistics considerations.

Separating the system into a number of CIs during development normally requires that a separate specification be written for each CI. This provides additional technical detail about the expected performance of the CI (over and above the requirements specified in the system, or higher-level CI, specification). This practice also provides additional technical management insight into the evolution of the CI's design. Each CI will have its performance and interface requirements documented in a separate Development (Hardware Configuration Items (HWCIs)) or Requirements (Computer Software Items (CSCIs)) Specification (MIL-STD-490). During the development of the design, each CI is normally subjected to a series of design reviews to assess the contractor's evolving design information against the requirements in the CI specification. When the development and testing of each CI is complete, it is normally subjected to a separate Functional Configuration Audit to verify compliance of the design with the CI specification performance requirements. And, when a deliverable unit of the CI with the operational configuration has been produced, each CI is normally subjected to a separate Physical Configuration Audit to verify the accuracy of the detail design documentation and to take control of the detail design. However, while it might seem to be a good practice to have a large number of CIs during development, the CI selection process must be a trade-off between the desire to manage configuration very closely at lower levels and the availability of adequate resources (primarily people and money) to do the job properly.

Once into the Production phase, and continuing on through the Operations and Support phase, additional configuration items are identified. By definition, any repairable item designated for separate procurement is called a configuration item. However, these "configuration items" are usually not documented with separate MIL-STD-490 specifications (like the CIs selected for development); rather, the drawings/technical data packages that define their design constitute their "specifications". Likewise, these "configuration items" are not subjected to separate design reviews or configuration audits at this point in the program; their evolving design was addressed as a part of the reviews and audits conducted for the higher-level CIs during the development process. However, the design of these "configuration items" must be controlled just as rigorously as the design of the higher-level CIs; to do otherwise would lead to logistics problems.

11.3   Baseline Management

The configuration of a system or CI consists of a Baseline Configuration (specifications, drawings and other configuration documentation) plus all changes that were approved subsequent to the approval of that particular Baseline. Two key points of the baseline concept are that there must be a *contractually-recognized* (baselined) and documented initial statement of requirements and that, once baselined, the current status of the configuration can be related back to those initial requirements. Baselines may be established at any time in a program when it makes sense to define a contractually-binding technical departure point for control of future changes in performance and design; however, there are certain points in the evolution of the system and CI design when it is normally prudent to establish a specific type of baseline. Three types of technical baselines are normally established over the life cycle of a system: the functional baseline, the allocated baseline and the product baseline. The following paragraphs discuss when the various baselines should be established in accordance with configuration management policy; however, the exact time of establishment of each baseline may vary from program to program.

The Functional Baseline will be a product of the initial conceptual effort (i.e. Concept Exploration and Definition or Demonstration and Validation phase effort, depending on the acquisition strategy used). It is normally defined using a Systems Specification for large acquisitions programs. (Smaller, single- item acquisition programs will normally use a Prime Item Development Specification instead.) This baseline defines top-level performance and interface requirements/constraints for the system (or the item) and serves as the contractually binding technical description of the expected system. The functional baseline will normally be established not later than the completion of System Design Review (SDR). This baseline, plus approved changes to the baseline, constitute the current approved functional configuration documentation.

An Allocated Baseline is established for each configuration item. It normally consists of a development specification (for hardware CIs) or a requirements specification (for software CIs) which defines the functional and interface requirements/constraints for that CI. The CI requirements are allocated from, and traceable to, the requirements in the System Specification/ Functional Baseline. However, you should understand that the CI's specification also expands upon the system specification requirements, providing additional detail requirements to supplement the very brief requirements for that item contained in the system specification. Thus, the allocated baselines might be better understood if you think of them as the expanded functional baseline prepared in a set of separate volumes (specifications), one for each CI. The allocated baseline for the CI provides a more detailed, contractually-binding requirements basis for the development of the detail design of a configuration item that will satisfy the requirements of the System Specification.

The Allocated Baseline will normally be established for the CIs at the start of the Engineering and Manufacturing Development (E&MD) phase, although it may be established during Demonstration and Validation if the acquisition strategy dictates. In any event, this baseline will normally be established before the PDR for the hardware CI and must be established not later than completion of the PDR. For computer software CIs, the allocated baseline must be established not later than completion of the Software Specification Review for the CSCI. [Please note that some programs will be developed under the Clear Accountability in Design (CAID) approach; for these programs, the CI specifications are generated at the normal time, but they are informally approved at the above milestones and are not baselined until after the completion of the FCA for the system (using the baselined system specification). This approach will provide more flexibility in making changes to the CI requirements without the need for engineering change proposals, but it will also limit the extent of contractual definition of the requirements that the system components must meet and of the verifications that will be used to demonstrate that their performance requirements have been met.

The development and requirements specifications that make up the Allocated Baselines for the CIs are normally (CAID programs excepted) authenticated and contractually invoked in the E&MD phase contract. During E&MD, the program office utilizes the Preliminary Design Reviews (PDRs) and the Critical Design Reviews (CDRs) to evaluate progressively more detailed design information about the CIs. The basis for determining the success of the contractor's effort to date is the requirements in the Allocated Baseline for that CI. When the verification (qualification testing) of the CI has been completed, the Allocated Baseline for that CI is the contractual basis for its FCA; failure of the design to achieve one or more of the required performance elements will result in the establishment of an FCA Action Item for the contractor to correct the discrepancy. Thus, the development and requirements specifications, as contractually-binding baselines, play an important role in the evaluation of the contractor's development effort and in the validation of the resulting CI design. CAID programs will accomplish these reviews at the system level, using the contractually-binding system specification.

The Product Baseline is defined using detailed design documentation for each CI. This documentation normally includes a product specification for the CI plus referenced (or included) drawings and related lists, source code listings, and software design documents. In addition, some CIs will also incorporate "program-unique" process and material specifications into their product baselines. (These specifications are normally used when a special, non- standard process or material has been developed specifically for use with this CI and when the use of that exact process or material is critical to achievement of the required CI performance.) The Product Baseline defines the exact design which the government wants delivered for use in the operational inventory. It provides the contractual basis for the acceptance of the production units of the CI; it also provides the basis for the entire logistics support system (spares, manuals, support equipment and software) for the CI. If CM has been performed properly up to this point, establishment of the Product Baseline will also signify that this CI design meets its parent requirements in the Allocated and Functional Baselines. The Product Baseline must be established not later than completion of the Physical Configuration Audit (PCA) for that CI. [CAID programs are expected to follow this same procedure.]

Once the program office authenticates the specification and establishes the baseline, those specification requirements for the CI and/or system constitute a formal contractual agreement on the required characteristics for the CI/system. Formal Configuration Control procedures commence at this point for the documentation that has just been "baselined". Changes to a baseline have contractual implications, so these procedures insure that the complete impact of the proposed change is presented and considered, and that the benefits to the government of the proposed change are considered, before a decision to approve or disapprove the change is made.

11.4   Configuration Management Functions

As mentioned earlier, the four functions of configuration management are Configuration Identification, Configuration Control, Configuration Audits, and Configuration Status Accounting. The importance and emphasis of each function depends on the size and nature of the acquisition program and on the phase in the life cycle. The CM director will organize the office in a way that will accomplish the CM functions needed for each phase of the system life cycle most effectively. The two most common organization structures are "project-oriented", where the configuration manager has the responsibility for all four CM functions for a particular project/program, and "function-oriented", where the configuration manager has the responsibility for one of the four CM functions for all projects/programs. This section discusses how each of these functions is accomplished.

11.5   Configuration Identification

The Configuration Identification function begins during the earliest phases of the acquisition process and continues throughout the life cycle. It includes the responsibility for ensuring that the following activities are accomplished as a part of the program:

- The selection of the configuration items to be managed by the program office during development.

- The selection of the appropriate type of specification and/or other configuration documentation to be used to document the functional and physical characteristics of each CI.

- The establishment of the contractual technical baselines.

- The establishment (by obtaining or issuing) of the identification numbers for the configuration items, their component parts and assemblies, and their related documentation.

SELECTING CONFIGURATION ITEMS

The discussion of the need for configuration items and of the factors to be considered in selecting configuration items was provided earlier.

SPECIFICATIONS

The baselines for the system and its configuration items are primarily defined in specifications. MIL-STD-490 establishes the format and content requirements for specifications prepared to define the functional and physical configuration of program-unique items, processes and materials. Its purpose is to prescribe uniform practices for preparing specifications, to ensure that all essential requirements are included, and to aid the program office and the contractor(s) in the use and analysis of specification content. MIL-STD-490 defines five types of specifications to document functional/performance requirements and product/physical characteristics at appropriate times in the system life cycle:

- System Specification.

- Development (or Requirements) Specification.

- Product Specification.

- Process Specification.

- Material Specification.

During concept development and exploration, the government develops a System Requirements Document (SRD) that incorporates all of the top-level requirements to be met by the new system; it is derived from the users' Mission Needs Statement (MNS) and other material developed through related studies and analyses. The configuration management group, if one exists this early in the program, will normally be required to maintain the current version of the SRD, even though it is never formally baselined. Part of the configuration identification process accomplished by the contractor or the program office will involve the conversion of the SRD into a draft System Specification that establishes the performance requirements, design constraints, and support requirements and the associated verifications for the system. This draft will form a key part of the Request For Proposal (RFP) for the demonstration and validation phase. The System Specification should conform to the requirements of MIL-STD-490, Appendix A. There should be frequent communication and careful coordination between the CM office and the Engineering offices during this period.

The System Specification released with the Demonstration/Validation RFP will often be an incomplete statement of the system requirements. Therefore, the source selection will include the negotiation of the contents of the final System Specification; contract award for the Demonstration/Validation phase may include the baselining of the System Specification, although many programs delay the establishment of the baseline until later in this phase. If the completion of the system specification is not possible during source selection, or if the program decides to delay the establishment of the functional baseline, the contract should contain a task requiring the contractor to complete the System Specification during the demonstration/validation phase. The System Specification should be complete by the end of the demonstration/validation phase, at the very latest.

The demonstration/validation phase contract should also require the contractor to prepare draft Development Specifications and Requirements Specifications for all CIs that are being designed specifically for use as part of this system. The contents of these specifications are generated as a part of the systems engineering process and are prepared according to the appropriate Appendixes of MIL-STD-490. Depending on the level of complexity, and the level of assembly of the CI, the requirements for the CI may be documented in a prime item, critical item, non-complex item, or software specification. MIL-STD-490 provides guidance for deciding which of these subtypes of specification would be most appropriate. As in the development of the System Specification, configuration management will be involved in the generation of these specifications, so there should be close coordination between the CM office and Engineering during the demonstration/validation phase. The CI Development (and Requirements) Specifications are normally prepared by the contractor and are reviewed and approved/contractually incorporated by the program office. Although the draft specifications will normally be prepared during the demonstration/validation phase, it is common to delay the approval/contractual incorporation of the final specifications until the early part of the E&MD phase. This would require a task in the E&MD contract for the finalization of the specifications. As mentioned earlier, this approval would normally occur not later that the completion of the SSR for software CIs and of the PDR for hardware CIs. CAID programs would normally "informally approve" the final draft CI specifications by these milestones, with the authentication waiting until after the system-level FCA.

The E&MD phase contract should also require the contractor to prepare draft Product Specifications for all CIs that are being designed specifically (and for some privately-developed items that need to be carefully controlled specifically) for use as part of this system. The contents of these specifications will be generated as a part of the systems engineering process and will be prepared according to the appropriate Appendixes of MIL-STD-490. Most hardware CIs will have their detail design documented by a product fabrication specification, which gives the government control over the detail design documentation. A few hardware CIs, usually privately-developed items which will be maintained by the supplier, will be documented in a product function specification, which gives the government control over the external package and the performance of the CI but does not provide control over the detail design. Software CIs will have their detail design defined in software product specifications. These product specifications are normally approved at, and the related CI product baseline established following, the Physical Configuration Audit. For software CIs, this tasking would commonly be a part of the E&MD contract; for hardware CIs, this tasking is normally a part of the Production contract.

For hardware CIs, there are options available that relate to the contractual status of the requirements in the development specification throughout the remainder of the life cycle for the CI. For many programs, the development and product fabrication specifications are prepared as separate documents with different identification numbers. This approach is used mainly when the development specification is required for the development phase of the program, but it will not be required in the production phase as we buy additional quantities of production units from the original contractor or from other suppliers; only the product fabrication specification will be needed to buy those production units. However, if there is a need to require compliance with the detailed performance requirements in the development specification by the original contractor or other suppliers throughout the remainder of the life cycle, then the "Two Part Specification" approach, as covered in MIL-STD-490, should be utilized. Using this approach, the development specification is controlled as a separate document until the product fabrication specification is approved/contractually incorporated; at that time the two documents are combined into a single document with a single identification number. Any production or spares contract will invoke the complete specification such that "the development specification (requirements) remains alive during the life of the hardware CI as the complete statement of performance requirements. Proposed design changes must be evaluated against both the product fabrication and the development parts of the specification." [MIL-STD-490]

As a part of the product baseline, some CIs may require program-unique process (e.g., the curing process for a solid propellant) and/or material (e.g., the solid propellant) specifications to completely define their characteristics. For each of these critical processes and materials, it is determined to be essential that the government have a document to precisely define its attributes and that we specifically address any proposed engineering changes against the process/material (as defined in the specification). Tasking for the generation of these specifications is normally a part of the E&MD phase contract; baselining may be accomplished in the E&MD or in the production phase. [On almost all programs, hundreds of processes and materials are listed on the drawings and other manufacturing documentation used to produce the CI; those standard processes do not require program-unique specifications. Control of changes to those processes and materials is exercised through the government's control of the design documentation (e.g., the drawings) on which they are listed.]

Specifications can be updated by either a change or a revision. A change is accomplished as the result of an approved Engineering Change Proposal, with the contractor issuing, and the government approving, a Specification Change Notice (SCN) and attached specification change pages. A revision is accomplished by the contractor, who generates and distributes a completely revised specification incorporating all previously approved changes (SCNs); this effort is required periodically throughout the program, normally as directed by the government contracting officer.

ESTABLISHING BASELINES

The discussion of the need for baselines and the normal (and CAID-related) timing for the establishment of the baselines was provided earlier.

IDENTIFICATION NUMBERING

The configuration identification process also includes obtaining or issuing identification numbers for the CIs and their documentation. Such numbers include specification and drawing numbers, part numbers, serial numbers, computer program identification numbers, national stock numbers, and nomenclatures. Some of these identifiers are generated and controlled by the contractor; others are issued and controlled by the government. The identification numbers provide us with a "shorthand" to utilize in referring to the items and the documentation; instead of using a string of several words, a single alpha-numeric designator suffices. (Requirements for the format and procedures for the issue of these numbers are contained in several military standards.) They are also used in much of the program documentation to refer to the items, their component parts, and the related documentation. However, if the contractor must obtain these numbers for your program, the contract must incorporate the appropriate tasking from these military standards.

Nomenclature is one of the designators which will be very important to your program. Nomenclatures allow us to maintain a relatively unchanging base number for a large number of functionally similar, but sometimes physically different, units of a particular major element of the system. It is especially important in setting up and maintaining the supply system for our operational inventory, but it is also necessary to maintain consistency between the items that we are buying and their related documentation. As the design of the system evolves, and as configuration items and their component assemblies are identified, the contractor must be tasked in the contract to submit official requests for basic nomenclature to the DoD Control Point for those items. The nomenclature will be affixed to the documentation and to the nameplates for the items, as they are prepared, so it is essential that the official nomenclature be used. MIL-STD-1812 provides specific instructions on the procedures to be followed in obtaining the assignment of an official nomenclature, and it specifically states that "A [nomenclature] shall be assigned by the development activity to identify a development equipment, group, or unit after first obtaining a basic assignment from the DoD Control Point." It also states, "The appearance of nomenclature for items covered by this standard in invitations for bids, contracts, specifications, drawings, and associated documents shall not constitute official assignment."

INTERFACE REQUIREMENTS IDENTIFICATION AND CONTROL

On all programs, the government will identify the contractually binding interface requirements in the functional and allocated baselines. MIL-STD-490 specifies that system, prime item development, software requirements, and (software) interface requirements specifications should contain requirements about interfaces. Once the specification is approved and the baseline established, these requirements can only be changed using a Class I engineering change proposal. However, looking at the "big picture" of all of the elements of the system being designed concurrently, the interfaces defined in these specifications constitute a very small percent of all of the interfaces that have to be defined and controlled by the contractor during the development portion of the program.

When the program office deals with a prime contractor who in turn has a number of subcontractors, the prime has the responsibility for defining and controlling all requirements below those specified by the government baselines. This responsibility includes the assignment and control of interfaces to be met by the subcontractors. This process is handled through the normal relationships that exist between parties to a contract.

However, when the government contracts separately with two or more prime contractors for development and production of elements that will be a part of the same system, this normal contractual relationship between the parties to the contract does not exist; the two (or more) contractors involved do not have contracts with each other. Under these circumstances, the program office must task each of the primes, in their contracts, to participate in an Interface Control Working Group (ICWG). MIL-STD-973 defines the basic requirements for such interface control activity; MIL-STD-499 defines the system engineering process, which includes the definition and management of the interfaces. The focus of the ICWG activity is on reaching agreements between the contractors about the interfaces that are not defined in the established baselines but that must be defined and controlled among the associate contractors in order for them to design their elements of the system. As problems arise, and changes may need to be made to the ICWG-established interfaces, the ICWG provides the forum for the review of the proposed changes and for deciding the way that the changes can be accommodated. In most cases, this means that one or more of the associate contractors will update their interface control drawing/document and provide it to the other affected ICWG members. In a few cases, however, this may mean the submittal of an engineering change proposal to the program office by one or more of the contractors to incorporate the newly defined interface into one or more of the contractor's specifications so that it can no longer be changed without government direction (and money).

11.6 Design Reviews

The design reviews are utilized throughout the development of a new system/CI to evaluate the contractor's technical progress. It is the intent of each review to assess whether the contractor has adequately completed the work so far, through the evaluation of design documentation expected to be completed by that point in the development process. Because the design evolution is usually related to certain points in the development schedule of the system, the design reviews are often considered to be activities that take place at a certain point in time in the program. While this is not totally untrue, the emphasis of the design review is focused on a specific package of design documentation. Thus, the purpose of each design review should be viewed much more as a review of a specific increment of documentation (at the appropriate point in time when it becomes available) rather than as a review, at a specific point in time, of whatever documentation might be available at that time.

The design reviews are a part of the system engineering process, as defined in MIL-STD-499. As such, they are the responsibility of the engineers; however, configuration management is usually affected by the results of the design reviews and must be involved in their accomplishment. Each of the series of reviews identified in MIL-STD-499 has a specific purpose, as briefly outlined below:

A System Requirements Review (SRR), or a series of SRRs, is used to evaluate the information in the draft System Specification to assure that it adequately specifies the requirements and constraints deriving from the Mission Need Statement that was the basis for the program.

A System Functional Review (SFR) is conducted as the final event in the development and finalization of the system specification. The closure of this process should be the approval/contractual incorporation (baselining) of the system specification. However, the SFR also looks at the draft specifications that have been prepared for each major configuration item selected to be managed separately as a part of the development process. The intent of the SFR is to assure that the requirements and verifications allocated (and expanded) into the draft CI specifications flow from requirements in the system (or higher-level CI) specifications. All of the requirements should be traceable back to the requirements in the higher-level specifications. They must also be deemed necessary and sufficient to serve as the basis for the development of the design of the CI.

A Software Specification Review (SSR) is conducted for each software configuration item selected to be managed separately as a part of the development process. The intent of the SSR is to evaluate the requirements and verifications allocated (and expanded) into the software CI requirements specification from the system (or higher-level CI) specifications. All of the requirements should be traceable back to the requirements in the higher-level specifications. They also must be necessary and sufficient to provide the basis for the development of the design of the CSCI. The closure of this evaluation process should be the approval/contractual incorporation (baselining) of the requirements specification for the CSCI.

The Preliminary Design Review (PDR) is conducted for each CI, and for the system, to evaluate the functional design of the CI and the selection of the necessary requirements from the requirements documentation for the various functions of the CI. If the functional design or the CI specification is determined to be inadequate, perhaps because it does not adequately address the system specification requirements, the contractor will be required to correct the situation, to the program office's satisfaction, before proceeding with the detail design process in that area. In some cases, the functional design work to date may have proven that a certain system requirement is unachieveable, in which case the correction will involve the submittal of an engineering change proposal to revise the requirement. The closure of this evaluation process should be the approval/contractual incorporation (baselining) of the development specification for the hardware CI and the authorization for the contractor to proceed with the detail design process. (The baselining may be delayed if the system is being developed using the CAID approach.)

The Critical Design Review (CDR) is conducted for each CI, and for the system, to evaluate the detail design of the system or CI against the contractual requirements in its baselined configuration documentation (most commonly, its specification). If the detail design is determined to be inadequate, perhaps because it does not adequately address one of the specification requirements, the contractor will be required to correct the situation, to the program office's satisfaction, before proceeding with building the hardware test units or with the generation of the software executable code. In some cases, the detail design work to date may have proven that a certain requirement is unachieveable, in which case the correction will involve the submittal of an engineering change proposal to revise the requirement.

The Test Readiness Review (TRR) is conducted for each software configuration item to assess the status of the software code. It is conducted after the completion of the coding of the CSCI executable code and after the informal testing at the Computer Software Unit and the Computer Software Component level has been completed. It's purpose is to review the results of the informal testing to verify that the CSCI is ready to begin the CSCI-level verification inspection process; it is also intended to review the adequacy and completeness of the detail testing procedures that will be used to conduct the verifications.

The CM Office will be concerned about the specifications being reviewed at the SRR, SFR, SSR and PDR; they will be used to establish the baselines for the system and the CIs and will be the basis for configuration change control throughout the development of the system and CIs. CM will also be concerned about discrepancies discovered at the PDR and CDR, since these may lead to the submittal of Class I engineering change proposals against the system and/or CI specifications.

11.7   Configuration Audits

Configuration Audits comprise the capstone technical events in the development of the system and configuration items. After CDR (and TRR for software), the contractor proceeds with a verification program to check the performance characteristics of the CI or the system. Data are gathered as a part of the conduct of the examinations, demonstrations, and tests and are analyzed and compiled into various reports. These verification results must be checked to assure that the design has achieved all of the specified requirements; a Functional Configuration Audit (FCA) is normally used to accomplish this task. Likewise, before the detail design is placed under contractual control, various documentation and process checks must be completed to assure that it is accurate and ready to be placed under formal government control; a Physical Configuration Audit (PCA) is normally used to accomplish this task.

FUNCTIONAL CONFIGURATION AUDIT (FCA)

The FCA evaluates the degree to which the CI's (or system's) performance complies with the requirements and constraints specified in its baselined specification. Theoretically, the CI (and system) should meet (or exceed) all of the requirements while operating within the constraints defined in the specification. Appropriate verifications (tests, demonstrations, examinations, and analyses) have been accomplished based on the Quality Assurance provisions specified in Section 4 of the specification; the resulting data is reviewed as a part of the FCA to verify the performance.

The FCA should include a formal presentation, by the contractor, about the test results and overall findings for the CI/system. The presentation should identify all requirements that were not met along with proposed solutions, an account of ECPs that were incorporated and tested, and a general discussion of the entire system development and test effort. As a part of the FCA session, the contractor and the government should identify all residual tasks (and agree on the scope of the effort and the suspenses for their completion within the current contract price) required to close out the development of the CI/system.

As a part of the FCA, the exact configuration of the CI/system that has been tested must be documented. The draft product specification is compared to the inspected articles to ensure that the contractor has accurately documented their physical configuration for which the inspection data was obtained. At the PCA, this design documentation will be compared to the design documentation that defines the item the contractor is delivering to the government; any differences will require further testing (and a "mini-FCA" as a part of the PCA) to verify that they also meet the "performance" specification requirements.

PHYSICAL CONFIGURATION AUDIT (PCA)

The PCA is the formal examination of the "as built" configuration of a CI against its technical documentation to ensure its validity before using it to establish the product baseline. To accomplish this, the PCA will include a detailed comparison of the engineering drawings to the fabricated hardware or a comparison of the computer program source code listings contained in the software product specification to a like listing from the deliverable software medium. Since the production configuration is likely to be different than the verification inspection (prototype) configuration, the PCA for a hardware CI typically occurs with the availability of the initial production unit of the CI, although there may be convincing reasons why a different unit might be selected for the PCA. For software, the timing of the PCA is more flexible. It can be accomplished simultaneously with the software FCA; it is more likely to be accomplished after the verification testing of the integrated hardware and software has been completed.

To be sure that the contractor can control the incorporation of future changes into the design documentation, the contractor's engineering release system is reviewed by tracking selected ECPs through the system. The draft Part II specification will be reviewed to assure that the performance requirements it contains are sufficient to verify adequate performance in each production unit as a part of the production acceptance tests. In a related activity, the acceptance test procedures and results will be reviewed to assure that the acceptance testing is sufficiently rigorous to identify any units that do not meet the performance requirements stated in the product specification. The PCA will also involve the review of the DD Form 250 to be sure that it accurately reflects the exact configuration of the deliverable unit that is being offered for acceptance as a result of the PCA.

MIL-STD-973 includes a number of sample certification sheets that can be used to accurately document the exact status of major portions of the

FCA and PCA efforts. Most programs utilize these certifications as a part of the official minutes of the audits.

11.8   Configuration Control

The Configuration Control function regulates all changes to the baselined configuration documentation. The proposed change to a baselined specification, or referenced technical document, is called an Engineering Change Proposal (ECP). The objective of Configuration Control is to assess the complete impact of the proposed change on all CIs and associated areas (e.g., logistics), to expedite the processing and incorporation of beneficial changes, and to prevent the acceptance of marginal or unnecessary changes. Necessary or beneficial changes are defined in MIL-STD-973 as those that:

1. Correct deficiencies.

2. Add or modify interface or interoperability requirements.

3. Make a significant and measurable effectiveness change in operational or logistics support requirements.

4. Effect substantial life cycle cost savings, or,

5. Prevent slippage in schedule or allow desired schedule changes.

(In the Air Force, configuration control has seen expanded responsibilities, under the name of Change Control, as it has been applied to contract documents such as the Statement of Work, the Contract Data Requirements List, or other contractually-binding task-type documentation. In this context, the change to a non-technical contract document is called a Task Change Proposal (TCP) or Contract Change Proposal (CCP).)

Developing the package of information completely documenting the impacts of proposed changes is an expensive activity for the contractor; likewise, the process of evaluating, revising, negotiating and incorporating changes is expensive activity for both the contractor and the SPO. To minimize this expense and to ensure that configuration control makes the most productive use of available resources, the Configuration Control function will often prescribe the use of (brief) preliminary ECPs, Advanced Change/Study Notices (ACSNs), and pre- ECP Technical Interchange Meetings (TIMs) in the ECP processing cycle. These provide a means for screening candidate ideas for ECPs; using them, we can eliminate an unnecessary change or select a particular alternative approach to a desired change, before a significant investment in resources is made by either the contractor or SPO. Either the contractor or the SPO can originate an ACSN (AFSC

Form 223 or similar). [A completed ACSN (and a brief Preliminary ECP) is usually no more than three pages long, whereas the formal change proposal may be 25-50 pages long and will include a formal cost proposal which is expensive to develop.] The ACSN describes the need for the change, presents one or more alternatives to accomplish the change, and estimates the impact of the change on system performance, schedule and life cycle costs. For routine priority change proposals, the CDRL will often prohibit the contractor from starting work on the preparation of a formal change proposal package prior to approval of an ACSN by the Program Office. Similarly, between the time an ACSN is approved and the ECP is submitted, the Program Office may utilize TIMs between the appropriate technical people on both sides to agree to the key aspects of an ECP before it is finalized and submitted. If only a few candidate changes are eliminated or improved through these means, the pay-back in resources saved (manpower/manhours and money, primarily) will be well worth the small amount of extra time required to process the routine changes.

ECPs (and CCPs) constitute a major drain of program funding reserves. They usually require additional effort to be accomplished by the contractor with related additional costs to be charged to the contract. There have been many cases where the accumulated costs of all ECPs for a program are a major portion of the final total cost of the contract, especially during development. There are a number of documented cases where the cost of the changes added to the original development contract actually exceeded the original contract cost. Since the monies available to fund the changes are in limited supply, it is critical that the CM office, in conjunction with the program control office, maintain estimates of the cost impacts of all outstanding ECPs and (TCPs/CCPs). "Outstanding" includes all those that have been approved but are not yet formally incorporated in the contract, those that have been received by the SPO but not yet approved, and those known to be in preparation (or which are expected to be prepared) but which have not yet been received. These projections should be briefed to the program manager and other program office functional managers on a regular basis (many programs do this at the start of each CCB) so that they are better able to see "the BIG picture" of the change activity for the program and will be better able to prioritize the changes being processed. This is a critical contribution the CM office can make in tying the technical status of the acquisition program to the business management requirements.

ESTABLISHING CONTROL

Formal Configuration Change Control procedures are instituted concurrently with the establishment of the functional baseline for the program. They are continued throughout the life cycle and include control of the documentation comprising the allocated and product baselines for each CI as those baselines are established later in the program. The contractor(s) and the program office follow MIL-STD-973 to prepare and review all proposed changes to the baseline(s). MIL-STD-973 requires that the ECP include a description of all known effects of the proposed change, including impacts on areas such as logistics support, training, numbers of personnel. The ECP must also provide the information about the exact changes that will have to be made to the documentation defining the Functional, Allocated, or Product baselines.

MIL-STD-973 provides criteria for the classification of engineering changes. Class I engineering changes require a change to the baselined documentation which also affects the functional and/or physical interchangeability of the old and new configurations, usually to the extent that there will be additional development and testing of the new configuration and updates to logistics support elements for the CI. Class II engineering changes, which are applicable only to the product baseline documentation, require a change to the baselined detail design documentation, but they do not affect the functional or physical interchangeability of the old and new parts and usually have no impact on the logistics support elements. Class I engineering changes require the preparation of a Class I ECP, in accordance with MIL-STD-973, and require formal approval by the program office. Class II engineering changes are usually submitted in the contractor's format and only require review and concurrence, by the government plant representative office, that they are valid Class II changes.

An engineering change is used when the redesign is intended to be acquired as the new, preferred configuration for the remainder of the acquisition of the CI. However, situations often arise on the program which lead to the deliverable unit having a configuration other than the preferred one specified in the currently approved configuration documentation. If the situation is discovered before the contractor begins final assembly of the deliverable unit(s) which will contain the discrepant (FAR calls them "nonconforming") part, the contractor would request a deviation from the currently approved configuration documentation requirements for the unit(s) affected. However, if a similar situation arises during the actual assembly or the acceptance testing of the deliverable unit, the contractor would request a waiver for the affected unit. In either case, the approval of the request by the program office means that the limited number of units having the different configuration will be used in the field without special procedures being required and without any need for subsequent action to retrofit those specific units. MIL-STD-973 includes the requirements for the contents and the processing of Requests for Deviations and Waivers.

CONFIGURATION CONTROL BOARD (CCB)

DoDI 5000.2 requires that each program establish a CCB for the purpose of reviewing proposed changes and of advising the Program Manager about those changes. The CCB is the official joint command/agency body chartered to decide on all Class I ECPs and on major and critical deviations and waivers. (It may also be used to make decisions on ACSNs and on TCPs/CCPs.) As the person responsible for the overall program, the program manager is responsible for the CCB decisions; for major programs, the program manager is usually the designated CCB chairperson. However, other program-related duties often consume so much of that individual's time that an alternate chairperson is usually designated. That may be the deputy program manager or the chief of Projects or Program Control or Engineering or Configuration Management. For smaller programs, a higher-level (normally field grade) supervisor responsible for a number of small programs being managed by individual (company grade) program managers will normally be designated as the CCB chairperson.

In order to provide the needed advisory capability for the CCB chairperson, the CCB must be comprised of individuals having knowledge about the various aspects of the program. As such, the CCB members are normally the top-level (and/or the most experienced) managers from each of the functional offices in the program office. This includes logistics, program control, contracting, manufacturing, engineering, configuration management, test, projects, using activity, training activity and similar functional representatives. Other members are included as appropriate to the program; these might include representatives from other DoD services (on joint service programs) and representatives from other countries (on multinational programs). The main point is to provide the CCB chairperson (the decision maker) with the most complete, comprehensive input possible concerning each change so that the decision will be based on a complete impact analysis.

CCB membership for a program should be formalized through formal designation of the chair person and members; various administrative options are available for this designation, but it is most commonly accomplished using a letter signed by the program manager or using a set of official administrative orders. In order for a person to sign the CCB Directive, they must be an official member (or alternate member) designated on the CCB orders. All of the organizations listed as regular members are expected to:

- Be represented at every CCB meeting;

- Be knowledgeable concerning the changes being considered, at least from their functional perspective;

- Participate in the CCB meeting by making their positions known to the chairperson; and,

- Sign the CCBD and note their agreement or disagreement with the decision.

It is important to note that DoDI 5000.2 states that the CCB is not a voting board – only the Chairperson makes the decision. The Configuration Management office also supplies the Secretariat (administrative) function which schedules and organizes the meeting, distributes the ECPs before the meeting, records the conduct of business at the meeting, records the CCB decision on a CCB Directive (CCBD), and delivers the CCBD to the Contracts Office so they can begin their negotiation and contract modification activities.

The CM office should publish procedures governing CCBs so that all representatives understand how the Board will operate and what their specific responsibilities are. The procedures should define target processing times for the ECPs within the program office; MIL-STD-973 provides the processing time requirements for arriving at the technical decision on the change.

- Generally, ROUTINE ECPs should meet a CCB no later than four weeks after receipt, although some situations later in the life cycle may dictate a longer processing time. ECPs that were preceded by an ACSN (or a formal TIM) should require no more than three weeks if they are reasonably consistent with technical agreements made during the ACSN (or TIM) phase. The only surprise in a routine change should be the cost or schedule impacts, although these aspects of the change should be addressed as estimates in the preliminary stages of change definition to avoid major surprises. A routine change will normally include a firm price proposal that can be negotiated and included in the contract along with the technical change.

- An URGENT ECP should be scheduled to meet the CCB within one week of receipt. An EMERGENCY ECP should be scheduled to meet the CCB not later than the day after it is received. Because of the urgency with which they must be prepared, reviewed, and approved, these ECPs will often contain an incomplete technical package and are required to contain only a "not-to- exceed" price estimate. Urgent/Emergency ECPs should be discouraged and only accepted when absolutely required. If approved, they represent an authorized, but not fully defined, technical change and cost increase; as such, there is usually some "wasted" technical effort and cost because of the "haste" with which the change had to be authorized. When they do occur and are approved, the CCB should assure that the change proposal includes a schedule for receipt of a formal ECP technical package and firm price proposal.

11.9   Configuration Status Accounting

Configuration Status Accounting (CSA) refers to the process of keeping track of all information relating to the baselined configuration documentation for the program. Traditionally, most people think of CSA in terms of paper reports or computerized data bases provided by a contractor once the program has proceeded into production and operation of the system. However, CSA is much more than that; it is the *information infrastructure* that allows the configuration management (hence the technical management) of the program to be effectively accomplished. It can be as simple as a pad of paper on a configuration manager's desk listing the ECPs currently being reviewed by the program office or as complex as the detailed parts list for a specific serial- numbered unit in the operational inventory. It is important to understand that CSA is independent of the medium on which the information is stored and of the activity (contractor or government) maintaining the information. CSA requires that information be maintained (and available when needed) about:

- The documentation (e.g., specifications, drawings, software listings);

- The identification numbers used for the documents and the items (e.g., specification numbers, drawing numbers, nomenclatures, part numbers, CPIN numbers);

- The status of action items established as a part of a configuration audit;

- The changes being processed (from receipt through contractual incorporation);

- The approved changes (CIs and serial numbers affected) and their implementation (e.g., production line incorporation, delivery of new spares, update of the ATE software, delivery of retrofit kits); and

- The configuration of the items being used in the operational inventory (on delivery and as a result of changes due to maintenance or retrofit/modification).

CSA starts very early in the program, usually with the establishment of the first baseline, and continues throughout the life of the system. Up until the end of the Demonstration/Validation phase of the program, the program office usually maintains any needed CSA information, primarily about the specification(s) and about engineering changes. As a part of the E&MD phase effort, the contractor is usually tasked to establish the information system for documentation, parts, identification numbers, and change implementation while the program office will continue to track the internal processing of the proposed changes. During the Production phase, the contractor is tasked to maintain information about the documentation, the identification numbers, the exact (by part number) configuration of each unit being delivered to the government, and the change implementation; the program office will maintain information about the internal processing of changes; the support activity will normally keep track of the current configuration of all units delivered and in the inventory. When production is completed, CSA during the Operational and Support phase is accomplished by the support activity; they are responsible for the maintenance and update of ALL the information already noted, even though these update/maintenance tasks were accomplished by the contractor or the program office in earlier phases of the program. Even after the item is retired from the inventory and placed in storage, CSA must continue; normally, it only requires maintenance of information about the actual parts in each unit in storage and of the related documentation. There have been numerous systems which have been withdrawn from storage for modification and further use in a new mission or as a low-cost weapon for our allies; the up-to-date information has been vital to the success of these efforts. Only when the item has been completely removed from DoD control (normally by disposal and destruction/recycling) can the CSA for that item (and unit) be terminated.

MODIFICATION MANAGEMENT

A modification is defined as a configuration change to a delivered configuration item that revises its functional and/or physical characteristics. Modifications can be made during the production phase of the system life cycle, but most occur during the operations and support phase. Some modifications are temporary; they are intended to be used to modify a unit(s) for testing purposes or for accomplishment of a special short-term mission; these modifications are intended to be removed by a demodification that returns the affected unit(s) to its original configuration and capabilities. On the other hand, major modifications are intended to effect a permanent change to the configuration and capabilities of the unit(s) affected. These modifications are used to correct system deficiencies, to counter new threats or otherwise increase operational capability, to lower life cycle costs, to extend the system's useful life, or to remove obsolete capabilities. AFR 57-4 provides the basic policy for the modification programs.

Effective modification management has most of the same attributes that effective configuration change control exhibits. It requires a disciplined CM system that:

        a. Identifies all functional and physical requirements that are affected, and all configuration documentation that must be changed,

        b. Develops a comprehensive ECP, and provides for a full CCB review and decision,

        c. Provides a timely and accurate accounting of the current configuration status of all operational items, and/or those in production, that will receive the modification, and

        d. conducts formal design reviews and configuration audits for large or otherwise critical modifications.

11.10 Summary

Configuration Management is an important tool that the PM must use to adequately define and control the technical side of an acquisition program (the TECHNICAL CONTRACT). CM mandates the acquisition and baselining of documentation to be used to define the technical capabilities/design for the hardware and software being acquired. CM mandates the verification that the established requirements have been met and that the documentation to be used for production is accurate. CM mandates the control of changes to the approved baselines which have been established to define the technical requirements for the system; the changes must include the complete summary of all impacts of any proposed change. CM mandates the availability of information about the documentation, the changes, and the operational units in the inventory. Configuration management, at the bottom line, provides the means of acquiring interchangeable items that can be effectively and efficiently supported throughout their life cycle.

References

[1] MIL-STD-490A, "Program-unique Specifications, Preparation of," A Version, 4 June 1985.

[2] MIL-STD-499A, "Systems Engineering," 1 May 1974.

[3] MIL-STD-973, "Configuration Management, 1 December 1992.

[4] MIL-STD-1812, "Type Designation, Assignment and Method for Obtaining," 28 February 1991.

[5] DoD Instruction 5000.2, "Defense Acquisition Management Policies and Procedures," Office of the Secretary of Defense, Under Secretary of Defense for Acquisition, 23 February 1991.

Chapter 12

DATA MANAGEMENT

12.1 Overview of Data Management

INTRODUCTION

Now that we have discussed how each technological alternative or weapon system is analyzed, how the alternatives are supported, and how the tracking of technological characteristics and changes to the critical items in the weapon system is accomplished, it is easy to see why a program manager's job is never done. There is so much information to sift through before a decision can be made, and so many documents that are generated that contain information necessary for the Government's monitoring of the status of a program. This data is managed according to policy identified in Department of Defense Instruction 5000.2. Without a comprehensive data management system, inefficiencies in the acquisition system would generate unnecessary duplication, insufficient types, and untimely receipt of data submissions. This chapter introduces the basic terminology used in data management and the tools available to get data on contract so that procurement of data is managed efficiently and effectively. Some of the data activities that typically occur in the system program office are addressed in tailoring the data requirements to your program.

DEFINITIONS AND TERMINOLOGY

Data management is the function that governs and controls the selection, generation, preparation, acquisition, and use of data from contractors. First, we will take a look at general terminology used in data management.

Data: Data is defined by AFR 310-1, Management of Contractor Data, as recorded information, regardless of form or characteristic. It includes all the administrative, management, financial, scientific, engineering, and logistics information and documentation required for delivery from the contractors. Data can be thought of as being either technical or nontechnical in nature. DoD Standard 963, Preparation of DIDs, distinguishes technical data as Type I and non-technical data as Type II. Type III data is one-time use data and can be either technical or nontechnical. The overall guidance for data is found in DoDI 5000.2.

In addition to these three types of data, it is important to know that our data is ordered under different Contract Line Item Numbers (CLINs) depending upon the type of data it is. The Technical Data Package (TDP), which includes all of our technical drawings, will comprise one of the CLINs. Technical Manuals (TM), which includes our tech orders, comprises a second CLIN, while any other data, called OTHER, comprises our third CLIN.

Data is acquired for two primary purposes: (1) Information feedback from the contractor for program management, control, and decision making; and (2) information needed to manage, operate, and support the system, such as specifications, technical manuals, and engineering drawings.

The following terms are key concepts that need to be understood by today's acquisition project officer. Data has become very expensive in relation to other components of the weapon system. One of the reasons for this high cost is the tremendous amount of data called out as deliverables in the contract that is deemed necessary for managing the weapons system. Sometimes that data is not really needed or could be tailored down, eliminating non-productive data requirements, so that the data could be delivered with less restrictions.

Acquisition Management Systems and Data Requirements Control List (AMSDL): The Acquisition Management Systems and Data Requirements Control List (AMSDL), DoD 5010.12-L, is an index which identifies acquisition management systems, source documents, and data item descriptions (DIDs) which have a standard application. The DoD Computer-aided Acquisition and Logistics Support (CALS) Program Office controls now the AMSDL and the processing for approval of data item descriptions included in it, and processes requests for new standard data items to the Office of Management and Budget (OMB). An approved data item is given an OMB control number, with an expiration date, as required by the Paperwork Reduction Act (Public Law 96-511).

Data Item Description: A data item description (DID) is a completed DD Form 1664, Data Item Description, that defines the data content, preparation instructions, format, intended use and recommended distribution of data that might be required of a contractor. The DID might be viewed as a specification for data to be generated and delivered as a result of a DoD contract. There are three types of data item descriptions: standard data item description, tailored data item description, and one-time data item description. It is interesting to note that there is not a DID for tech orders, but a contract requirements document TM 86-01, that is used to get the required data.

Standard Data Item Description: A Standard Data Item Description is a data item that has been approved for general use. It is listed in the AMSDL, and published and distributed to the military services, Federal agencies, and to subscribing DoD contractors by the Naval Publications and Forms Center in Philadelphia, Pennsylvania.

Tailored Data Item Description: A Tailored Data Item Description is a standard item data that exceeds the requirement for information and must be tailored downward, or diminished, to meet the specific requirement. Tailoring may only be accomplished to:

1. Accept contractor's format.

2. Reduce the scope through deletion of words, paragraphs, or sections.

Note that data item descriptions can only be tailored down, and never be tailored up; tailoring up is against the law.

One-time Data Item Description: A One-time Data Item Description is a data item that is developed when a data requirement cannot be met by use or tailoring of a standard item, or by combining submittals of multiple-tailored standard data items. A one-time DID has a 6-month time-frame to be put on a contract. Once on contract, it is valid for the life of that contract. It is assigned an OT DID number from an approved block of OT numbers (as of this writing). The one-time data item description is typically controlled by data management focal points at product centers. (NOTE: One-time DIDs are not listed in the AMSDL, nor are they printed and distributed to the field. The contract they are part of is the only place they occur. The DD Form 1664, detailing the one-time DID, is incorporated in its entirety into the particular contract.)

Because of the limited use, a one-time DID is used only to meet the needs of the contract for which it is developed. If future uses are anticipated, the one-time DID is forwarded, through data management channels, to DDMO for consideration by OMB for approval as a new standard DID.

Contract Data Requirements List: The Contract Data Requirements List (CDRL) is a list of data requirements that is authorized for a specific procurement and is made a part of the contract. This list is comprised of a series of DD Forms 1423 (individual CDRL forms) which contain the DID identification numbers and delivery instructions. The CDRL is one of the two places in the contract where you can establish a requirement for the contractor to deliver data. The other way is by specific contract clauses, formerly called the general provisions section, which brings in Federal Acquisition Clauses applicable to your contract. Both the CDRL and the clauses require OMB control over the data being collected, regulated by the Paperwork Reduction Act (Public Law 96-511).

Contract Data Requirements List data may be thought of as data directly linked to statement of work (SOW) tasks and is managed by the Data Management Officer (DMO). Detailed instructions for completing the DD Form 1423, Contract Data Requirements List, are found in AFR 310-1. Data required by FAR clauses most often deals with the sound business aspects of contracting and is managed by the Contracting Officer.

DATA CALL AND DATA REQUIREMENTS REVIEW BOARD

The Program Manager is responsible for acquiring the contract data necessary to manage all aspects of the program/project. A Data Management Officer (DMO) is usually appointed to assist the Program Manager in this task. The process of identifying and acquiring the required data begins with the concept exploration and definition phase and continues throughout the entire life cycle. For each contract to be issued (usually for each phase of the program), the formal process begins when the DMO issues a data call.

Data Call: The data call is usually a letter that describes the planned program and asks functional managers to identify and justify their data requirements for that contract. When responding to the data call, the Air Force uses a special data ordering form; AF Form 585, Contractor Data Requirement Substantiation, explained in the AFR 310-1. This form is similar to the DD Form 1423. It also contains blocks to justify the need for the data ordered; show impact for not ordering the data; and provide the name, phone number, and office symbol of the individual ordering the data. The data call is sent not only to the different functional offices within the program office, but also to all commands and agencies involved in an acquisition.

When the data call letter is issued, functional managers have their personnel find the sections they are concerned with in the Statement of Work (SOW) and have them determine which SOW paragraphs tasks cause data to be generated. Once the personnel have identified the paragraphs, they next identify the types of data that should be generated, and then look in the AMSDL to determine which DIDs will meet their data needs. Once this is accomplished, they identify their information needs on the AF Form 585. They review the DIDs to assure that they provide the needed data, and then the data item descriptions are tailored to eliminate unnecessary requirements. This tailoring process is like special-ordering a new car, with the DID being similar to the dealer's catalog. You need to clarify some options by identifying the specific exterior and interior colors. In other cases, you have redundant options (such as tire types and sizes), and you identify those you want. In many cases, there are options you don't want, and so you tell the salesperson you don't want them. In this way, you order the car you want without paying for the cost of options that you do not want.

If a standard data item description cannot be identified or tailored to meet requirements, a one-time data item description is generated on a DD Form 1664, using the guidance in DoD Standard 963, Preparation of DIDs. These requirements are sent to the DMO, usually on AF Form 585, which identifies the required data item description, delivery instructions, and provides the justification for the acquisition of data item.

Data Requirements Review Board: Once inputs are received from the data call, the DMO compiles all the requirements and attempts to eliminate redundant data items. The next step is for the Program Manager to review the requirements and the associated justification. The review is usually done in a meeting called a Data Requirements Review Board (DRRB). This board is normally comprised of representatives from the functional areas having significant data requirements. The board does not vote; rather, it recommends. The chairperson of the DRRB makes the final decision. (The Program Manager may chair the DRRB, or delegate this role to the Data Management Officer). Based on this meeting, the Program Manager decides which data items are included in the Request for Proposal going out for the contractors to respond to.

Several DMOs are called in to provide technical evaluations to the negotiating team commenting on the contractor's proposed data preparation and pricing effort. The DMO, in conjunction with the contracting office, then finalizes the desired contract data requirements list (CDRL), negotiates it with the contractor, and makes it a part of the contract.

After contract award, the DMO is responsible for tracking and timely delivery of all of the CDRL data on the contract. If the contractor is late on delivery or if the data delivered is deficient (including having restrictive markings not authorized by the contract), the DMO, through the Contracting Officer, can use the FAR clause entitled, "Technical Data–Withholding of Payment," to withhold up to 10 percent of the contract price from the contractor in order to press for the late/deficient data.

If the Government is late on granting approvals to the contractor on any data submitted requiring Government approval, the DMO aids the Program Manager in speeding up the Government approval process, reducing Government-caused contract schedule slippages.

DEFERRED DATA

Now, we will take a look at the four types of deferred data: deferred ordering, deferred delivery, deferred requisitioning, and the data accession list.

Deferred Ordering: Deferred ordering is explained in the Federal Acquisition Regulation (FAR), Part 27, which deals with the Acquisition of Technical Data and Computer Software. This type of data is expensive to prepare in the required form, and costly to update and maintain. Examples of such data are engineering drawings, computer software, and technical manuals. Deferred ordering of data is applied contractually, via a FAR clause, and provides for delaying the ordering of technical data generated in performing the contract until a need for the data can be established and the data requirements can be specifically identified for delivery under the contract.

The typical situation where deferred ordering applies is an undefined design where one is "pushing the state of the art." The data which has been deferred is later defined, the actual order for the data is placed contractually, and delivery is scheduled. The only charges for this data are for collection, formatting, printing/electronic transmitting, and computer tape copying. The charges for all engineering efforts were previously been made against the statement of work tasks (as opposed to data preparation expense). There is a 3-year rule built into the FAR deferred ordering clause which sets the limitation on when deferred ordering can occur.

Deferred Delivery: Deferred delivery is also explained in FAR, and is applied contractually, via a FAR Clause. Deferred delivery refers to those situations where receipt of the technical data or computer software by the Government would be premature for adequate storing and handling. The typical situation, where deferred delivery applies is when the Government is unable to store, update, or retrieve data. Costs for deferred delivery data include only the cost to collect, format, print, or make a tape. There is a 2-year rule built into the FAR deferred delivery clause which sets the limitation on when deferred delivery can occur.

Deferred Requisitioning: Deferred requisitioning is also done contractually; however, no FAR clauses apply. Instead, the Contracting Officer develops a clause describing the instant contracting situation. The terms, prices, and delivery schedule are established up front; however, the data arrives later. There are no year rules; the Contracting Officer could negotiate a time frame to fit the situation. Typical situations where deferred requisitioning apply include lack of final design, hardware un- deployed, and Government in-house inability to manage the data. What are the advantages of deferred requisitioning? The Government does not have to store, update, or retrieve the evolving design-related data. The contractor does this and charges for the additional service. You can see that these deferred data techniques incur additional cost.

Data Accession List: The Data Accession List (DAL) technique employs a DID for the listing of non-CDRL data which evolves over the life of the contract. The contractor creates a library of this non-CDRL data. The DID only gets the list of what is in the library. The Government buying office can order data deliverables from this list at any time. There is not approval of any item from the list, nor is there any say on format or content. The Government is charged for copying costs and for the storing and retrieving from the contractor's data library. There is, obviously, some additional cost to set up a DAL mechanism on a contract, but it is another way to have access to practically all the remaining data which is not formally ordered on CDRLs.

Reprocurement Data: Another type of data that is sometimes necessary in system procurement is the acquisition of reprocurement data. The acquisition of reprocurement data is necessary when the Air Force requires follow-on competitive reprocurement of an item, or of spare parts for the item. The reprocurement data may contain specifications, drawings, and/or special processes and other data concerning:

1. Physical description of the item such as drawings (paper, microfilm, computer aided design digitized on computer tape or laser disc) and specifications (all types).

2. Parts/materials/processes including vendor and subcontractor lists.

3. Operations and test criteria including software and automatic or special test equipment.

4. Manufacturing, specifically tooling and computer-aided manufacturing; quality assurance documentation; and manufacturing sheets and routings.

5. Proprietary documentation including licensing.

6. Other requirements, especially test requirements documents.

Engineering Data Management Officer: Meeting minimal data requirements while satisfying program needs is definitely a current issue. The public's awareness of spare parts overpricing problems in the Department of Defense has been a growing concern for all the services. The Air Force has taken a unique approach toward solving the problem of achieving viable reprocurement data packages by designating a new position of Engineering Data Management Officer (EDMO), as a specialized data person to support the Program Manager in acquiring viable reprocurement data packages. The EDMO role is evolving, and is being developed at all Air Force Materiel Command (AFMC) Product Centers and Air Logistic Centers (ALCs). The EDMO, involved in the whole engineering drawing process acquisition, ensures contractor contract compliance in the preparation of engineering drawings and reprocurement data packages and coordinates with the AFMC EDMO to ensure timely delivery of the engineering data at the lead ALC for the system.

THE PAPERWORK REDUCTION ACT OF 1980

Public Law 96-511, states that only authorized data published in the AMSDL or the FAR is required to be formally submitted to the Government. Authorized data means the data request document, or DID, as written or as tailored down, a one-time DID. In other words, contractors are not required to provide any CDRL data as a formal data submittal which is illegally tailored (i.e., the DID requests information above and beyond the published, official data item description). This means that Government personnel have to be very careful in generating data requirements and putting them on a subsequent contract.

DD Form 1423: The DD Form 1423 is the form that becomes part of the Contract Data Requirements List (CDRL), and is a very important part of the contract that you will have to be able to understand to get a better handle on the data you should be receiving. Let's look at some of the more important parts of the DD Form 1423 so when you see one on your program you'll be better able to read it.

Block 1: Data Item Number: The Data Item Number is the CDRL sequence number. This number is important to know if you are looking through the statement of work (SOW), see a CDRL sequence number listed, and are interested in knowing what the data item description represents.

Block 2: Title of Data Item: The title of the data item is same as the title entered in item 1 of the DD Form 1664.

Block 4: Authority (Data Acquisition Document Number): This number is the same as item 2 of the DD Form 1664 and will include a "/t" at the end of the data acquisition document number (e.g. DI- MCCR-80237/t) if the DID has been tailored.

Block 5: Contract Reference: To help identify the data item authorized by the Block 4 Authority, the applicable document and paragraph numbers should be entered that identifies the task or requirement from which the data flows (contract clause, statement of work paragraph, etc.).

Block 6: Requiring Office: The requiring office is the activity responsible for advising on the technical adequacy of the data.

Block 7: Specific Requirements; Specific requirements may be needed for inspection/acceptance of the data.

Block 8: Approval Code: If an "A" is in this block, it is a critical data item requiring specific advanced written approval prior to distribution of the final data item. This data requires submission of a preliminary draft prior to publication of a final document. When a preliminary draft is required, the Remarks Block, Block 16, should show length of time allowed for government approval or disapproval and subsequent turn-around time for the contractor to resubmit the data after government approval or disapproval has been issued. Block 16 should also indicate the extent of the approval requirements.

Block 9: Distribution Statement Required: Although out of order, this block is left until last to describe because it is relatively new. The Scientific and Technical Information (STINFO) program is to ensure that all scientific and technical information (STINFO) generated under Air Force research, development, test and evaluation programs is made available and utilized to the maximum extent possible. In order to do this without jeopardizing our Country's security, it is necessary to place the proper distribution statement on our data.

AFR 83-3 establishes both policy and procedures for applying distribution statements. We will give a brief synopsis of each of the categories here, with the hope that you will refer to AFR 83-3 when you are building your data requests. Category A is an unlimited release to the public. Category B is limited to release to government agencies. Category C limits release to government agencies and their contractors. Category D is limited to release to DoD offices and their contractors. Category E is for release to DoD Components only. Category F is released only as directed, and is normally classified data. The final category is X and deals with export control issues.

Distribution marking is but one facet of STINFO, and there are activities that should be performed in each program office in conjunction with the STINFO program. Each product and logistics center has a STINFO representative that needs to be consulted in planning for STINFO activities.

Block 12: Date of First Submission: This block should indicate the year/month/day of the first submission. If the delivery of the data depends on the initiation of a specific event or milestone, this event or milestone should be identified. If you do not know the contract start date, the number of days the data is due after the contract start should be given. If this data is not known or needs more clarification, explain in the Remarks Block.

Block 13: Date of Subsequent Submission: If you need the data to be submitted more than once, then the dates of subsequent submission should be entered in this block. If a draft was required, the date the final document needs to be submitted by is entered also in this block. If this event would cause the data list to become classified, leave this space blank.

Block 14: Distribution: In this block, you need to identify each addressee and give the number of copies to be received by each one. If possible, use office symbols, contractor initials, or DoD Handbook H-4 code numbers and command initials; in this case, attach a list explaining these codes. If reproducible copies are required, give the type of reproduction requested here, or in the Remarks Block (i.e., offset, mat, negative, floppy disk, magnetic tape, etc.). If the data are not actually to be delivered to the government or to associate contractors, the appropriate instructions should be given.

Block 16: Remarks: This space is used to explain any tailored features of the DID, any additional information for Blocks 1 through 15, and any resubmittal schedule or special conditions involved in updating data submitted for government approval.

12.2 Discussion

As can be seen from the description above, determining your data needs is something that requires a great amount of though, especially in light of the mandate to only purchase the absolute minimum data that we need. There are several things that need to be reiterated from the preceding discussion. First, you must go through each data item and tailor it to exactly what you need, remembering that you can only delete, not add. Contractor's format may be acceptable in many cases, or contractor's format with a matrix built into the DD 1423 showing where the contractor's format information meshes with the required data item information are two ways we can save money and still get the information that we need.

We need to insure that or contract references will actually generate the type of data that we are looking for, and that any contract tasking we have that generates data has the appropriate data item description tagged to it; we often miss in both instances.

We need to ensure that when we indicate that a document needs approval that we really need to review a draft prior to the document being released to the world. We can always require the contractor to resubmit a document that does not match the data item description or that has glaring deficiencies. Approval is probably necessary on specifications, but probably is not necessary for test reports. What we have to remember is that approval documents are more expensive to obtain, and should be held to the minimum that we need.

Determining the time of first, and subsequent submissions, is many times best left to tying the submissions to major milestones. By tying the data to major milestones, we have fewer headaches if the calendar schedule changes in terms of when to submit data. When we tie data to a milestone, we have to give consideration to how long before or after the milestone that we should get the data. Using a preliminary design review as an example, if we ask for the specifications too far in advance, we will not have to most current information. If we ask for the documents too close to the review, we will not have enough time to adequately review them to be prepared for the review. There has to be a balance that can only be determined by each program office.

When we determine our distribution list, we must take a close look at who needs to receive the data, and make sure that they get it, and seriously question requests from organizations that don't seem to need the distribution they have requested. Again, we usually miss out on both points. We also need to take a real hard look at the number of copies requested, and ask ourselves if it is vitally important that everyone have their own personal copy, or will several copies do for a large office.

Finally, we must realize that normally there is a clause at the beginning of the contract data requirements list that gives time frames that both the government and contractor have to review the data in. If we do not review the data in the specified time frame, normally 45 days, the contractor can assume that the document is satisfactory and we will not be able to comment on it. 45 days may seem like a long time, but it goes fast! The other thing that we are famous for doing is something many of our teachers and bosses do to us when we write something for them; that is, they review it one time, make corrections, review it a second time, make corrections, review it a third time, and make corrections, and so on. We can only review the contractors document once, so we need to do so thoroughly and indicate all changes that have to be made. If the corrections are inadequate, we can ask the contractor to fix those, but we can't enter additional comments about areas outside of those we already indicated need correction. This is something that happens frequently, and can make getting a document an almost endless process. One other effective technique is face to face reviews with the contractor if you are having substantial problems coming to an agreement on a document. The richness of the communication medium usually gets faster action than letters that can be misinterpreted.

Data is a necessary, expensive part of any contract. We must fully understand what our data needs are, what data we have on contract, when the data is due, and who should get the data. If we do the data function properly, we will be leaving a supportable program in our wake.

References

[1] DoDI 5000.2, "Defense Acquisition Management Policies and Procedures."

[2] DoD 5010.12-L, "Acquisition Management Systems and Data Requirements Control List (AMSDL)."

[3] DoD STD-963, "Data Item Descriptions (DID), Preparation of."

[4] AFR 310-1, "Management of Contractor Data."

[5] AFSCR 310-1, "Management of Contractor Data."

[6] AFLCR 310-1, "Management of Contractor Data."

Chapter 13

COMPUTER HARDWARE AND SOFTWARE MANAGEMENT

13.1 Introduction

Mission critical computer resources have become a major component in many of the weapon systems that we have recently developed, and are currently developing. The "smart" weapons employing the most recent computer technology demonstrated a decided advantage in the recent Desert Storm operation. Some modern aircraft have been likened to flying suites of computers surrounded by an airframe. In order to fully comprehend the field of mission-critical computer resources (MCCR) software support management, a manager must have a basic understanding of the relevant terms and concepts. This chapter explains the fundamental terms of computers, software, programming languages, and firmware for those with little or no computer background. The chapter is also intended to be a useful review for those experienced in computers and software. This chapter, however, does not go into great detail on any of these terms and concepts.

13.2 The Computer System

A computer may be defined as "a functional, programmable unit consisting of processing unit(s) and peripheral equipment that is controlled by internally-stored programs and which can perform substantial computations without human intervention." This definition applies to a wide range of computers from a hand-held, programmable calculator to a large mainframe computer with a room full of equipment.

An understanding of the terminology associated with computers and software is essential to understand Mission Critical Computer Resources (MCCR) management. Computers are computational devices which normally consist of four components: a central processing unit, secondary storage, input units, and output units. This grouping of components is sometimes referred to as a computer system. Software instructs the computer to do certain functions. Programming languages translate human language into binary pulses which a computer can then interpret and act upon.

COMPUTER HARDWARE

The four basic components of a computer: the central processing unit, secondary storage, input devices, and output devices will now be examined in greater detail.

Central Processing Unit (CPU): The CPU is the "heart" of the computer. It normally contains an arithmetic/logic unit (ALU), control unit, and primary storage. The control unit selects instructions from primary storage, interprets them, and tells other hardware components to execute them. The ALU performs all calculations and makes all comparisons or decisions necessary for manipulation of data from primary storage as directed by the control unit. Primary storage, or main memory, is used to hold the programs and data that the computer is currently processing. Primary storage is made up of random access memory (RAM) which is temporary, i.e., the contents are lost when the computer is turned off, and may be randomly, rather than sequentially, accessed. Random access permits quick access to any needed portion of RAM. In addition to RAM, many computers have read only memory (ROM) as part of the internal memory. This memory is unalterable and its contents are not lost when the power is off. It usually contains instructions that help the computer work.

Secondary Storage: Since the contents of primary storage are lost when the power is off and the cost is relatively high, secondary storage is required for long-term safekeeping of large amounts of computer data on all but the simplest computers. The most common examples of secondary storage are magnetic tapes and disks and optical disks. Less frequently used devices include magnetic drums, magnetic bubble memories, and charge-coupled devices. It is the responsibility of the CPU to initiate transfer of data to and from secondary storage as needed.

Input Devices: Input devices are used to transfer data into the computer system for processing. Common examples of input devices include the keyboard, optical scanner, bar code reader, and mouse. Less frequently used, some once commonly used, include the light pen, paper tape, and card readers.

Output Devices: Output units receive data from the CPU and prepare it for use by a computer user. Common examples of output devices are cathode ray tube (CRT) displays, printers, and plotters.

Special Components: Although all computers have the basic components just described, the characteristics of the components may differ markedly from the examples presented. One example is a computer embedded into a weapon system such as a B-1 bomber, where input units may be radars, antennas, or similar devices. Output units may include weapon release devices or flap settings. Secondary storage may consist of additional circuit cards. Different computer applications, therefore, often require different types of components.

COMPUTER SOFTWARE

A computer, in some ways, is like a 2-year old child; it must be told what to do. Also, like a child, a computer must be commanded in the right way or it will not respond. Worse yet, a computer does not understand everyday English; it only works in "computerese" which consists of electrical signals represented by binary numbers 1 and 0. Groups of these binary numbers are interpreted as instructions or data to be acted upon. Sets of related instructions are called computer programs, a key element in the definition of "software." Software may be defined as "computer programs, procedures, rules, and possibly associated documentation and data pertaining to the operation of a computer system."[1]

Firmware: Firmware is a special case of software in that it exhibits characteristics of both software and hardware. Firmware is defined as "computer programs and data loaded in a class of memory that cannot be dynamically modified by the computer during processing." This differentiates firmware from other storage media such as tapes or disks, as the software on these media is routinely modified during computer operations. A hardware device, normally a "chip" or integrated circuit, and a specially coded software program, sometimes called "intelligence," are combined in special equipment, such as a "burner." This equipment produces a read-only memory (ROM) device, a hardware device with the software embedded into it. There are a number of firmware devices that we employ on our weapon systems, such as Electronically Erasable Programmable Read Only Memory (EEPROM) and Ultra-Violet Programmable Read Only Memory (UVPROM) that give us the flexibility of placing the software on the computer chip as many times as we desire, allowing for software updates, without having to replace the chip.

Languages: As discussed previously, a problem exists when you try to tell a computer what to do. You want to give a command in English; for example: "I would like to go 50 miles from Dayton to Cincinnati in 1 hour. How fast must I travel?" The computer, which understands only binary numbers, 1s and 0s, cannot readily interpret this request directly. Therefore you must have your commands translated into the language the computer can comprehend. This introduces the concept of language translation, a concept where we can use different types of languages and have them translated into a language the computer can comprehend. Let's take a look at the types of languages we use on a computer, starting first with the language the machine comprehends directly.

Machine Language: A program consisting of groups of binary numbers can be executed directly by the computer. These groups of binary numbers are often called machine language, since they can be understood by the computer ("machine"). The programs are also referred to as "object programs" or "object code." Indeed, the earliest programmers had to write their programs in machine language because there was no alternative. However, machine language is not the language of choice for most programmers, except for special applications where direct communication with the machine is needed.

Assembly Language: The next step up the hierarchy of languages is assembly language. Assembly language uses sets of letters and numbers to give commands to a computer instead of groups of binary numbers. There is usually a one-to-one correspondence between assembly and machine instructions. A special computer program, called an assembler, translates assembly language programs to machine language programs. The commands used in assembly language, such as SUB (for "subtract"), STR (for "store"), ADD, etc., are certainly more English-like than machine language. However, writing an assembly language program requires some knowledge of the internal operations of a computer, and most programmers find assembly language cumbersome. Another disadvantage of assembly language is that it is not normally interchangeable between computers with different CPUs. An assembly language program for an IBM computer, for example, will not run on a Control Data Corporation (CDC) computer or other make.

Higher Order Languages: The next level up the chain of abstraction are Higher Order Languages (HOLs), a further step toward an English-like language. A programmer requires much less knowledge of the computer's internal workings to use a HOL than to use assembly languages. For example, FORTRAN, a scientific HOL, is well-known in this respect. Another advantage of most HOLs is that fewer statements are needed for a given program. One HOL statement usually replaces several assembly or machine instructions. However, HOLs do have certain disadvantages. Sophisticated programs called compilers translate HOLs to machine language. Many compilers are not as efficient as assemblers in that the machine language programs generated do not always make optimal use of the computer's storage (memory) or processing speed capacity. For these reasons, in applications where computer capacity is constrained, assembly or even machine language programs are still used. Common examples are portions of software programs for airborne weapon systems where speed of execution is often at a premium.

Modern programming technology has produced programs that are even more English-like than HOLs. These include such programs as spread sheets and word processors that allow a person with little or no knowledge of computers or HOLs to easily use a computer for certain applications. Some computer systems can even respond directly to a limited range of voice commands, making a direct translation from the English language to "computerese."

Source Programs: Important for both HOL and assembly language programs is the concept of source programs. A source program is a program written in any language, which must be assembled or compiled to be executed. This factor differentiates source programs from object programs, which are machine language programs generated by compilers and assemblers. For most applications it is important that both a source program and object program be retained. The source listing is needed for clarity and to make changes, while an object program is required to avoid having to recompile a program each time it is run on a computer.

Standard Higher Order Languages: A purported advantage of HOLs over assembly language programs is that they are portable; that is, they can be used on different computers without modification. In the early days of HOL, however, this advantage was slow to be realized. Many different HOLs were developed, each for a specific computer application. There was little or no commonality within the same HOL for different computers. For example, FORTRAN for an IBM computer had to be modified extensively to run on a CDC computer. It soon became evident that some sort of standardization among HOLs would be highly desirable, and the concept of a standard HOL evolved.

Standardization: Standardization of a language implies that the language is rigorously defined in terms of syntax (its grammar) and semantics (what it means). Any changes to the language must be rigorously controlled. The language compilers are also standardized because any compiler deviation constitutes a deviation from the language standard. Ideally, a standard HOL is completely portable among different computers; but, in reality, modifications often have to be made to a language for use on different computers.

The advantages of having a standard HOL include portability and reduced training times. Portability supports reusability of software–the same program can be used on several different computers or for different applications. Reduction in training time results in reduced costs and increased programmer productivity due to familiarity with one language. This can result in higher reliability and maintainability of software. There are some disadvantages of standardization. The main one is that special capabilities of a particular computer cannot always be exploited. Also, new ideas are not as easy to incorporate. For most applications, however, the advantages of standardization outweigh the disadvantages.

Standardization Authorities: The American National Standards Institute (ANSI) was the first agency to standardize HOLs. HOLs currently described by ANSI standards (ANSI-STDs) include COBOL and FORTRAN. The Institute of Electrical and Electronic Engineers (IEEE) standardized several languages. The Department of Defense (DoD) standardizes several languages, primarily for military use, such as JOVIAL. In addition, DoD has attempted to limit the total number of HOLs used for military use.

Common HOLs: DoD 3405.1, which has ben superseded by DoD 5000.2, listed nine approved HOLs: COBOL, C/ATLAS, FORTRAN, JOVIAL, CMS-2, SPL/1, BASIC, Pascal, and Ada. The general HOLS–FORTRAN, COBOL, C/ATLAS (and most other HOLs)–are not really geared to the real-time processing requirements of military weapon systems where even a split-second delay can be critical. But they can be used for applications discussed below.

The COBOL language is used extensively for business applications such as payroll and file management. It is very English-like and is relatively easy to learn. The current standard for COBOL is ANSI-STD-X3.23-1985.

C/ATLAS is a specialized language for automatic test equipment (ATE) test programs. In the past, the DoD has experienced problems with ATLAS because a kludge of different versions existed. Several different versions are still in existence, although IEEE-STD-716 ATLAS appears to be the current standard version of this language.

The FORTRAN language is frequently used in scientific and technical applications. It is ideal for coding complex scientific formulas. The current standard for FORTRAN is ANSI X3.9-1978. FORTRAN has been used for some military weapon systems such as the Precision Location Strike System (PLSS). Real-time suitable HOLs.

The remaining HOLs–JOVIAL, CMS-2, SPL/1, BASIC, PASCAL, and Ada–are better suited to real-time requirements and are approved for use by DoD. JOVIAL is one of the currently-approved languages suitable for real-time military applications. JOVIAL J-73, described in MIL-STD-1589C, is the current standard version used on many Air Force programs. Although the Air Force has used JOVIAL extensively, other services have used different languages such as CMS-2.

CMS-2 was a Navy Standard language similar to JOVIAL which has been used on many Naval programs. Eventually, JOVIAL and CMS-2 may no longer be DoD-approved HOLs, as they are projected for replacement by the Ada language.

SPL/1, like CMS-2, has been primarily used by the Navy for special processing applications. The language is controlled by the Navy. The Air Force used a different version of PL/I for programming the mission planning software at HQ SAC.

The (minimal) BASIC language is a favorite with neophyte programmers because of its simplicity. It has limited usefulness within DoD. The current BASIC Standard is ANSI X3.60-1978.

The PASCAL language contains special features which make it suitable for scientific and near-real time applications. PASCAL is a foundational language for Ada. The current PASCAL Standard is ANSI/IEEE 770X3.97-1983.

The Ada language is supposedly named after Augusta Ada Byron, Lord Byron's daughter and the first programmer. In 1983, DoD policy dictated that all defense mission critical application software would be written in Ada by 1984. In reality, this did not always happen because a full set of proven support tools for Ada has not always been available. However, the intent is to eventually make Ada the one standard language for these and other applications. DoDI 5000.2 directs the use of Ada for all DoD applications unless a waiver is granted. Ada's features are similar to those of PASCAL, and is supposedly superior in many aspects to other real-time languages. Ada strongly supports the use of modern software design practices and programming techniques which have been shown to enhance software development and support. The interested reader may wish to consult ANSI/MIL-STD-1815A or other related publications for additional information on Ada.

The use of one standard language such as Ada for all applications can have many benefits, as was previously discussed. It is improbable, however, that Ada will be the universal language, as there are specialized applications more suitable for languages such as COBOL, ATLAS, or FORTRAN. Also, it is conceivable that a language superior to Ada will make Ada obsolete in the future. Still, Ada will be the darling of DoD for sometime to come.

13.3 Types of Computer Resources

To fully understand the management of mission critical computer resources (MCCR), a manager must understand the peculiar aspects of MCCR that differentiate it from other types of computer resources. This section contains an explanation of the concept of MCCR, and the documents at various DoD management levels that affect MCCR management. Also included is a discussion on non-deliverable computer resources that present special challenges to MCCR managers.

MCCR is a relatively new term that can be best understood by a historical review. In the early seventies, computers and software for DoD weapon systems were receiving increased attention. A project called "Pacer Flash" was initiated to develop unique managerial procedures for computer resources. Some DoD computer and software directives existed, but they were mainly geared toward automatic data processing equipment (ADPE) applications such as payroll computation and scientific analysis. These ADPE directives were not really appropriate for the unique requirements of weapon systems computers and software. DoDD 5000.29, Management of Computer Resources in Major Defense Systems, the first DoD-level directive for computer resources, defined a new term, Embedded Computer Resources (ECR). ECR includes all computers, software, data, documentation, personnel, and supplies embedded within or integral to a weapon system. ECR were differentiated from ADPE in the way they were to be managed and the governing documents. At the Air Force level, the 800-series (weapons system) regulations governed management of ECR while the Air Force 300-series regulations governed management of ADPE.

With this separation of computer resources into two classes, problems soon arose. There were many "gray areas" which were neither "pure" ECR nor ADPE. For instance, cryptological and intelligence computer resources could be categorized as ECR because of their relationship to weapons systems, but they weren't really an "integral part" of a weapons system. Attempts to manage software as both ECR and ADPE did not succeed because of conflicts between 800-series and 300-series regulations. Adding to the confusion was a third type of computer resources, primarily for communications systems, which were managed by the Air Force 700-series (communications) regulations. It was apparent that something had to be done to rectify this situation.

The most recent solution, implemented by the Warner amendment to the Congressional Brooks bill in 1982, divided DoD computer resources into two classes: Information Systems Resources (ISR) and MCCR. ISR now includes ADPE and communications software. The Air Force 300-series regulations for ADPE have been incorporated into the Air Force 700-series regulations, which previously addressed only communications software. Meanwhile, MCCR includes additional resources to those included in the ECR definition. The definition of MCCR in the latest edition of AFR 700-4, Volume 2, Information Systems Acquisition and Major Automated Information Systems Review Requirements, includes the following five applications along with their respective sub-applications:

- Intelligence activities.

- Cryptological activities for national security.

- Command and control.

- Computer resources integral to a weapon system (i.e., ECR).

- Resources critical to military and intelligence missions.

This classification of computer resources into ISR and MCCR will not solve all problems. For example, there is controversy over whether robotics software and other nondeliverable computer resources (NDCR) used in DoD defense plants should be MCCR or ISR. Those advocating the MCCR label state that, since the NDCR performance is critical to successful weapon system production, they are MCCR per the fifth item of the definition. Others disagree, however, and believe that these resources are ISR. One way to avoid such controversies is by combining all computer resources into a single category. However, such a move would face stiff opposition.

13.4   MCCR Software Life Cycle

The software life cycle is described in DoD-STD-2167A, which includes a description of the software management principles for mission critical computer resources. These principles may also be applied to any type of software development. We have already discussed the system life cycle process, and software development is a subset of that life cycle. You will remember from Chapter 10, Systems Engineering, that we are defining system requirements throughout concept exploration and definition, and through demonstration and validation. Although some preliminary work will be accomplished in demonstration and validation, most of the software development activities occur during the engineering and manufacturing development phase. The eight phases of the software development process, as described in DoD–STD–2167A are:

1. SYSTEMS REQUIREMENTS AND DESIGN

The software development process begins with the allocation of system–level requirements of Hardware Configuration Items (HWCIs) and Computer Software Configuration Items (CSCIs). The System/Segment Design Document (SSDD) describes this allocation. After the SSDD is approved, CSCI-development can begin.

2. SOFTWARE REQUIREMENTS ANALYSIS

The purpose of this phase in the software development cycle is to define and analyze requirements of a Computer Software Configuration Item (CSCI), which is a software program that has been selected for formal configuration control. The requirements should be quantified and measurable, and must relate to the overall requirements of the weapon system on which the CSCI resides. Therefore, interface requirements for the specified CSCI to interact with other CSCIs and HWCIs are also established during this phase. The requirements for the computer software configuration item are documented in a Software Requirements Specification (SRS) and reviewed at a Software Specification Review, which normally occurs very early in engineering and manufacturing development. Successful completion of this review normally determines the allocated baseline for the software configuration item, although requirements analysis may continue on a diminished basis as the software design evolves.

3. PRELIMINARY DESIGN

Once requirements for a CSCI are established, the design process begins. During the Preliminary Design Phase, a design approach, including mathematical models, functional flows, and data flows, is developed. Trade-off studies are normally conducted to determine the best design approach. This is the phase of software development where the contractor defines the Computer Software Components (CSCs), which are a functional decomposition of the CSCI, much like the functional decomposition we discussed in systems engineering. An example of this would be to imagine having a CSCI from the Operational Flight Software of a missile where the software was supposed to control the missile, compute the missile trajectory, perform navigation functions, arm the warhead, and perform built in test. Any one of these previously listed functions would make a good candidate for a CSC under the operational flight software CSCI.

Software requirements are allocated to the CSC-level, and the contents and interface requirements for each CSC are defined. Critical Computer Software Units also may be designed. The results of activities during this phase are reviewed at a Preliminary Design Review, and the go ahead is given to proceed with the detailed design. The documented description of the CSCI design can be found in the Software Design Document, and is sometimes referred to as a "Developmental Configuration," which can serve as a basis for future design enhancements.

4. DETAILED DESIGN

The purpose of this phase is to refine the design approach by breaking out each CSC into Computer Software Units. The Computer Software Unit (CSU) is a further functional decomposition of a CSC, and is the lowest level of code that we manage. A CSU is typically 200-400 lines of source code, although a CSU may be considerably smaller using Ada, and normally represents a single function that has an input, a process, and an output. Using the example we started in the last section, let's choose the Navigation CSC. Some logical CSUs might be "Calculate a Turning Radius", or "Compute Speed", or "Update heading", or any of many more possibilities, depending on what the CSC was supposed to accomplish. The important thing to realize in this example is that we take a CSCI and functionally decompose it, just like we do the overall system in systems engineering, to get down to a level with a single, identifiable function.

The detailed design of the CSCI addresses both internal and external interfaces along with data bases to be used by elements of the CSCI. This design is normally recorded in a Programming Design Language (PDL) which may, or may not, bear a strong resemblance to the eventual source code. The results of activities during this phase are documented in an expanded Software Design Document, which still represents the developmental baseline, and are reviewed at a Critical Design Review which, if successful, gives the go-ahead to proceed to coding.

5. CODE AND UNIT TEST

Ideally, until this time, no operational software code has yet been written. During this phase, both source code is written and object code is generated for each CSU of a CSCI. A Software Development File (SDF) is kept on each CSU to record its implementation schedule, its requirements, its design, unit test plans, unit test results, the actual unit source code, and a history of modifications. It is through careful inspection of the Software Development Files that the government program manager and engineer can assure that the effort is being handled properly and has a schedule that will support program milestones. In addition to being coded, each CSU is also tested to ensure that it can perform its required subfunction.

6. COMPONENT INTEGRATION AND TESTING

During this phase, aggregates of coded CSUs are integrated and tested to ensure they can interact as CSCs. This phase often overlaps the Code and Unit Test Phase, since some CSUs may be ready for integration while others are still being coded. A Test Readiness Review is held at the end of this phase to determine whether a CSCI is ready for CSCI-level testing.

CSCI TESTING

The purpose of the final phase of the software development cycle is to test a fully implemented CSCI to ensure that it meets requirements. The results of this testing are used as a basis for formal acceptance of a CSCI by the Air Force as a completed product. All required documentation for the CSCI should be complete by the end of the phase.

8. SYSTEM TESTING

Although CSCI Testing completes the development cycle for a CSCI, the CSCI may still undergo changes as a result of system-level testing. A CSCI must successfully interface with other CIs, and problems in system-level testing may necessitate changes in a CSCI which has already been developed. Furthermore, many changes must be accomplished to each CSCI after system testing and the EMD phase are completed. With few exception, CSCIs undergo many reiterations of part or all of the software development cycle before the software is no longer used. These reiterations are often referred to as "software support." The totality of all software development and support activities for a CSCI constitute that CSCI's "life-cycle."

13.5   MCCR Development

While the life cycle just discussed sounds wonderful in theory, in reality many times it does not work that way. And, after all, we still haven't told you how we stuff all of that software into the computer. Typically, as some of the contractor's personnel are looking at the requirements, other personnel are already starting to make some design decisions, as well as deciding equipment needs. The computer software is not normally developed on the computer that it will eventually reside on. Normally, the software is developed on a large mainframe computer, and after being developed, is cross-compiled to the operating instructions that are found in the computer that it will eventually reside on.

When we go to the Software Specification Review, we must have done a thorough job of reviewing the software requirements specification because it becomes the document that we must live or die by in proving that our software does what it is supposed to. Many times the contractor is more interested in the designing and coding of the software than in making sure the requirements are well defined and understood, and we need to try and balance that.

In getting ready for the preliminary design review, we must make sure that our contractor has his facility ready to do our software work, and that he has employees that are knowledgeable about the coding language we are using. This is particularly critical with Ada, since it hasn't been around very long, and there aren't a lot of people familiar with it. Again, homework prior to the review in terms of requirements traceability will pay large dividends in confidence that the software being developed will do what it is supposed to do.

As we approach the critical design review, we will have found that the contractor has probably done a certain amount of coding, since the schedule we give him for development isn't normally very adequate. The contractor does this at his own risk, but is often times hesitant to change something that is already there, making our job more difficult. Traceability keeps cropping up as a topic, but we must perform a good requirements traceability prior to the critical design review to ensure that all of our requirements have been allocated to one or more units.

Moving into formal coding and unit test, we must pay particular attention to the software development files and the contractor's software library, where configuration of the evolving software is maintained. If library practices are sloppy, it will be difficult, if not impossible, to trace the evolution of the design to resolve potential problems. Part of this is keeping the documentation in the software development files up to date, and building a system to track the unit coding schedule to provide early warnings of problems.

As the contractor starts to actually code the software on the large mainframe computer, he will probably test the code on another mainframe computer that will model the computer the code will eventually reside in. Later in the process, he will have a working model of the actual computer the code will reside in, and perform further testing there. This often becomes a bottleneck in software development, especially if the computer provided is also developmental and subject to problems. Eventually, the contractor will possibly have to develop some simulation software to see if the code on the intended computer will worked in a modeled system. The final test, probably what will be used for qualification testing, will be the software in the intended computer, hooked up to a "hot mock-up" or interconnected pieces of the eventual system to show that all the pieces work together. This is an exciting and hectic time, especially as all of the various pieces of equipment come together.

Careful monitoring of the software development files can help provide confidence that things will work well together as we start to integrate them. One thing we must constantly be aware of is what the current sizing and timing of the software is. If the software is too slow, or if it is too big to fit in the computer memory, we have problems. Possible solutions are a bigger (memory) or faster computer. Other approaches may require trade–offs such as using assembly language for portions of the code to improve the execution speed with a corresponding decrease in maintainability.

As we approach formal qualification testing, we must ensure that validity of our lower level test results, and that adequate plans and procedures have been devised for qualification testing. Be aware that qualification testing is not as rigorous as lower level testing should be, its intent is to show that the software meets its requirements, we aren't testing to try and break it. If the more demanding testing wasn't done earlier, we won't see it now. We also have to plan for problems with qualification test, and start to develop plans for what set of regression tests, a repeat of previous tests, that we will have to perform to ensure that "fixes" were just that, and that they did not create additional errors.

Once through with qualification testing, we still have to perform a functional and physical configuration audit. You can see a functional audit, but what exists to perform a physical audit on? The code, in our case the source code. We must ensure that the source code matches the documentation in the design document to allow for better maintainability later on. We have to ensure that all problem reports have been resolved, all action items resolved, and that the contractor's software library is functioning properly so we get the proper version of all of the software units.

This has been a quick look at how we actually perform software development, with a few of the pitfalls to watch out for. An understanding of the theory versus the reality will help you have a better view of how software is developed on the program you work with. Good software development can be accomplished, but it requires a disciplined effort from the inception of the effort.

Mission critical computer resources are an area of growing importance to our weapon systems. To date, ignorance and fear has kept them from being managed properly, and this ignorance and fear does not seem to be diminishing at a very rapid rate. This lesson has introduced you to the tip of the iceberg of mission critical computer resources. Hopefully, if you find yourself assigned to this type of work, you'll avail yourself to DoD-STD- 2167A as well as the wealth of information currently available about software development and testing.

References

[1] Ferens, Daniel V., "Defense System Software Project Management," Air Force Institute of Technology, Wright-Patterson AFB OH, 1990.

[2] DoDI 5000.2, "Defense Acquisition Management Policies and Procedures."

[3] DoD Directive 3405.1, "Computer Programming Language Policy."

[4] DoD-STD-2167A, "Defense System Software Development."

[5] DoD-STD-2168, "Defense System Software Quality Program."

[6] AFR 800-14, "Life Cycle Management of Computer Resources in Systems."

Chapter 14

INTEGRATED LOGISTICS SUPPORT

Now that we have introduced the systems engineering process and the design reviews that allow the Government to monitor this process, how do we plan for the support of that technology being developed and ultimately produced as a weapons system? The answer is through integrated logistics support. This chapter addresses the concept of Integrated Logistics Support (ILS) and how it relates to Department of Defense acquisitions. Obviously, supportability of a weapons system is crucial to deployment of the weapons system. Logistic support planning begins as soon as the weapons system concept is defined and continues throughout the system's life. In this chapter we will expound on the ILS process and its elements.

14.1   Introduction

Integrated Logistics Support (ILS) has become increasingly important to the Air Force as our weapon systems have become more and more complex. This complexity in the weapon systems has made our systems more difficult and expensive to maintain. We have found that the most cost-effective approach is to design our systems with support in mind early in the systems acquisition process.

The systems acquisition process is a myriad of events which must be accomplished for the development, production, and deployment of a system. When viewed from a life cycle cost (LCC) perspective, systems go through sequential cost stages, and these stages are typically:

- Research, Development, Test and Evaluation (RDT&E).

- Acquisition (Production/Construction).

- Operation and Maintenance (O&M).

- Deactivation/Retirement.

When looking at the costs associated with each of these stages over an extended period of time, it has been discovered that the O&M cost of weapon systems has steadily risen to a point where they now consume more than 50 percent of the systems total LCC. This varies markedly when analyzing individual systems, but when viewed as an average across the DoD spectrum, it still represents a real concern. Figure 14.1 depicts the Life Cycle Costs over the life of a typical major weapon system, where some 10% of the total costs are used up RDT&E, 30% for production, and 60% for operations and support.

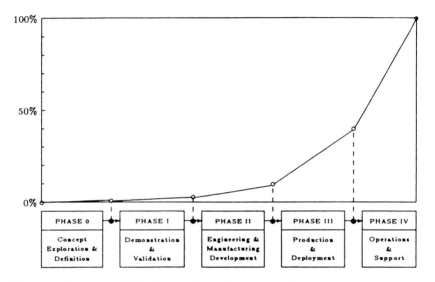

Fig. 14.1: Percent of Life-Cycle Costs for a typical major weapon system.

Further impacting system LCC is the timing of key program decisions such as: defining the operational scenario, establishing quantitative performance requirements, quantifying the number of systems to be fielded, specifying deployment locations, selecting a maintenance concept, etc. Studies have shown that by the end of systems concepts studies, 70 percent of the decisions defining total LCC have been made; by the end of the system definition, 85 percent have been made; and 95 percent by the end of Engineering and Manufacturing Development. This does not mean actual expenditures; it infers future spending has been committed because of the decisions that were made. It must be added that once these decisions are made, changing them at a later date could drastically impact program costs. It becomes necessary that changes be evaluated for cost versus benefit before the change is implemented.

Knowing that, on the average, the greatest amount of total LCC dollars are spent on O&M, which really equates to logistics support, and that the vast percentage of total LCC is committed early in the acquisition process, it becomes apparent logistics planning must begin at the front end of an acquisition program. The remainder of this discussion deals with the process of integrated logistics support (ILS).

Acquisition Logistics: In order to understand integrated logistics support (ILS) you need to be familiar with the definition of acquisition logistics. Acquisition logistics is defined as:

> The process of systematically identifying and assessing logistics alternatives, analyzing and resolving ILS deficiencies, and managing ILS throughout the acquisition process.[1]

Acquisition logistics, then, is a generic term identifying and describing the overall logistics function within which ILS is the predominant activity.

Integrated Logistics Support:. Department of Defense Instruction (DoDI) 5000.2 describes the ILS program as a disciplined, unified and iterative approach to the management and technical activities necessary to:[1]

1. Integrate support considerations into system equipment design.

2. Develop support requirements that are related consistently to readiness objectives, to design, and to each other.

3. Acquire the required support.

4. Provide the support during the operational phase at minimum cost.

Simply stated, the four functions identified in the definition of ILS mean that:

1. As systems are being developed, emphasis must be placed on designing-in, to the maximum extent feasible, those capabilities that improve and enhance logistics support.

2. As the logistics support package for a system is being planned, developed, and produced, a paramount concern must be to obtain the right combination of logistics elements that maximize system readiness at minimum life-cycle cost. It is not only necessary that each logistics element be optimized to the system it supports, but they must also be optimized to each other.

3. Deciding how the logistics support system must function and its composition is only part of the challenge. In most programs, industry plays the major role in system planning, design, and fielding. It is incumbent upon the program logisticians to translate logistics requirements into contractual requirements and then, more importantly, to ensure the requirements are met. Do not lose sight of the fact, however, that a significant portion of the logistics resource needed to support a system may come from existing DoD inventory. Therefore, equal attention and coordination must be given to the identification and acquisition of these resources through Government agencies.

4. After a system is fielded, the logistics support structure devised for it is put to the true test. It would be unrealistic to expect every facet of the logistics infrastructure established for a system to perform as planned. With this as an accepted given, follow-on actions must be pursued to correct deficiencies. In addition, you can expect systems to be modified sometime during their life cycle. Integrated Logistics Support planning does not end until a system is retired from the inventory, so the process of planning and implementing logistics support continually evolves.

What exactly is the definition telling us? First, it stipulates that the entire ILS process when used on a program must be diligently applied (disciplined). No two acquisition programs are alike and likewise, no two ILS programs in support of an acquisition are alike. However, all logistics elements, which are defined later in this chapter, must be evaluated for program applicability. When it has been determined what specific elements are necessary, they must be pursued methodically. Also, inherent in this definition is the uncompromising need for the ILS process to be performed and managed as a single entity (unified). The interrelationship of the ILS elements to each other is such that a change in one could greatly alter requirements in another. This impact could be in cost, schedule performance, design, or even the very need for an element may become questionable.

In order for the ILS program to be effective, it must be periodically and systematically reviewed and updated as the program progresses. Acquisition programs are extremely dynamic. Numerous factors, both in and out of program office control, change the way a program advances through the acquisition cycle. The results of internal trade-off studies, congressional or higher headquarters intervention, changes in requirements from the using command–all can be equally valid and necessary, but can also be equally devastating to the timely development of the logistics support structure for a system. All changes must be evaluated for their impact on the ILS elements and the other logistics processes involved in the ILS program.

From this, plans must be revised to incorporate the change with minimal disruption and expense to the program and to the systems future logistics support.

14.2   Integrated Logistics Support and the Life Cycle

Support costs for a weapon system are a major portion of a system's life cycle costs, and this lesson introduces the concept of how ILS relates to these life cycle costs. Integrated logistics support (ILS) begins when a DoD need is identified in a Mission Need Statement. ILS follows a weapon system from program inception until it is phased out of the USAF inventory. The program manager, who is responsible for ILS planning and management prior to Program Management Responsibility Transfer (PMRT), usually assigns ILS responsibility to the Deputy Program Manager for Logistics (DPML), and in some cases, to the Integrated Logistics Support Manager (ILSM). The ILSM usually works for the DPML and with complex, major programs, there may be hundreds of functional ILSMs working for one DPML. Some non-major programs may have an ILSM and no requirement for a DPML. In all cases, however, the ILS effort is the program manager's responsibility prior to PMRT.

INTEGRATED WEAPON SYSTEM MANAGEMENT

Based partially on the problems associated with PMRT, the Air Force decided to use a new process, when Air Force Materiel Command was formed, the Integrated Weapon System Management (IWSM) process (For details see Chapter 5). Integrated Weapon System Management is the process of managing eight critical sub-processes from the time a program is first conceived until the program is retired from service. One important part of the concept is the idea of having a single manager throughout the entire life cycle process, with the manager transferring to the Air Logistics Center with the program.

14.3   Integrated Logistics Support Documentation

There are two primary documents that support the Integrated Logistics Support planning; one prior to PMRT, the Integrated Logistics Support Plan, and one after PMRT, the Integrated Weapon System Master Plan.

Integrated Logistics Support Plan: The Integrated Logistics Support Plan includes all of the ILS panning prior to Program Management Responsibility Transfer. This plan is generally written by either the Deputy

Program Manager for Logistics, or some one he designates. The plan is generally started when a single concept is finally chosen for demonstration and validation, and is a living document until PMRT, requiring numerous revisions prior to that milestone.

Integrated Weapon System Master Plan: The Integrated Weapon System Master Plan is the 10 year plan developed by the System Program Director after PMRT. You might be asking yourself if 10 years is enough, and the answer is that the plan always extends 10 years in advance, being updated annually to include the next year.

With the formation of the new Air Force Materiel Command, it is uncertain as of yet what the plan will be called for logistics planning.

14.4   Logistics Support Analysis

The enormity of the logistics effort demands a structured technique to analyze, quantify, and integrate the totality of the logistics support resource requirements. The system DoD has employed to perform this analysis is logistics support analysis (LSA). Logistics Support Analysis (LSA) is simply a vehicle for implementing ILS requirements during the acquisition process. DoDI 5000.2 and MIL-STDS-1388-1A and 2A/2B stipulate more about LSA. Specifically, DoDI 5000.2 states:

1. The logistics support analysis process will be used to: (a) Develop and define supportability related design factors, and (b) Ensure the development of a fully integrated system support structure.

2. This process will incorporate, but not duplicate, analysis and data required by other functional disciplines.

3. The logistics support analysis record (LSAR) will be established for recording, processing, and reporting supportability and support data and will be used as the definitive source for this data.

It can be seen from the previous paragraph that Logistics Support Analysis (LSA) is the iterative analytical process within systems engineering that integrates logistic support requirements or constraints into the design of the hardware and software. It can also be seen that the Logistics Support Analysis Record (LSAR) is the medium through which task results and support source data are recorded to help develop and validate support capabilities and requirements.

14.5   Integrated Logistics Support Elements

The Integrated Logistics Support (ILS) elements subdivide the ILS program into manageable functional areas and disciplines. You must realize that it doesn't matter whether the program is a large one, like a new intercontinental ballistic missile (ICBM) or a small one, like a new helmet for pilots, all logistics elements must be evaluated for applicability to a program. For all practical purposes, what changes between large and small programs is the depth of effort to be performed in each element, even though both programs may have incorporated the same elements into their ILS planning.

Department of Defense has identified 10 elements in DoDI 5000.2 that represent the DoD perspective of logistics. These 10 elements include: maintenance planning; manpower and personnel; supply support; support equipment; technical data; training and training support; computer resources support; facilities; packaging, handling, storage, and transportation; and design interface.

MAINTENANCE PLANNING

Maintenance Planning evolves from the definition of systems operational planning and establishes maintenance concepts and requirements for the life of the system. It includes but is not limited to: levels of repair, repair times; testability requirements; support equipment needs; manpower skills; facilities; interservice, organic, and contractor mix of repair responsibilities; site activation; etc. It is this very element that establishes the baseline for planning, development, and acquisition of other logistics support elements. As you read the descriptions of the remaining ILS elements you should ask, "How would this logistics element be affected by the maintenance concept for the system?" You will soon realize the impact in nearly all cases is monumental. It goes without saying that the future holds interesting challenges for maintenance planners. Even now systems are being planned that require innovative maintenance approaches, for instance, fighter aircraft that will be launched and recovered on a highway rather than a permanent base, and mobile ICBMs in a nearly constant state of movement. Think about the maintenance possibilities for the orbital manned space station. The conventional two- or three-level maintenance concept must, by necessity, give way to a new dimension of maintenance support.

MANPOWER AND PERSONNEL

Manpower and Personnel involves the identification and acquisition of military and civilian personnel with the skills and grades required to operate, maintain, and support systems over the systems' lifetime. Early identification is essential. If the needed manpower is an additive requirement to existing manpower levels of an organization, a formalized process of identification and justification must be made to higher authority. Add to this the necessity to train these persons, new and existing, in their respective functions on the new system, and the seriousness of any delays in the accomplishment of this element becomes apparent. In the case of military requirements, manpower needs can, and in many cases do, ripple all the way back to recruiting quotas.

SUPPLY SUPPORT

Supply Support consists of all management actions, procedures, and techniques necessary to determine requirements to acquire, catalog, receive, store, transfer, issue and dispose of spares, repair parts, and supplies. In layman terms this means having the right spares, repair parts, and supplies available in the right quantities, at the right place at the right time. The process includes provisioning for initial support, as well as acquiring, distributing, and replenishing inventories. Keep in mind, an aircraft can be grounded just as quickly for not having the oil to put in the engine as it can for not having the engine.

SUPPORT EQUIPMENT

The Support Equipment (SE) element is made up of all equipment (mobile or fixed) required to support the operational and maintenance requirements of a system. This includes ground handling and maintenance equipment; tools, metrology and calibration equipment; and manual and automatic test equipment. During the acquisition of DoD systems we are expected to decrease the proliferation of support equipment into the DoD inventory by minimizing the development of new support equipment and giving more attention to the use of existing DoD or commercial equipment. Most programs are a mix of common and peculiar (commercial and new design) SE. Equal emphasis must be placed on the identification, funding, and acquisition of both. A program office may be making a serious mistake if they think the SE that is currently in the DoD inventory will be available when needed. It may well take longer to get some of the DoD inventory items than it would for new design items provided by the contractor. Another key point must be made; the availability of the prime system can be heavily influenced by the SE used for fault detection/isolation and repair. Much of the SE used today are repairable items themselves and therefore requires the timely development and fielding of a logistics support system for the SE as well. This means that SE also need maintenance plans, technical orders, spares, facilities, trained manpower, support equipment, etc. It should be obvious that if the support equipment isn't available because it cannot be repaired, the availability of the prime mission equipment could be affected.

TECHNICAL DATA

Technical Data represents recorded information, regardless of form or character (such as manuals and drawings), of scientific or technical nature. Computer programs and related software are not technical data; documentation of computer programs and related software is. Technical Orders (TO) and engineering drawings are the most expensive and probably the most important data acquisitions made in support of a system. It is the TOs that provide the instructions for operation and maintenance of a system. Without them it may be difficult, if not impossible, to operate and/or maintain the prime system and support equipment. Engineering drawings are crucial to a system's life cycle cost, allowing competitive reprocurement of spare and repair parts and the modification of systems, which in the long run should minimize the system's life cycle cost.

TRAINING AND TRAINING SUPPORT

Training and Training Support consists of the processes, procedures, techniques, training devices, and equipment used to train civilian and military personnel to operate and support a system. This includes individual and crew training, new equipment training, initial, formal, and on-the-job training. Though the greatest amount of training is accomplished just prior to the fielding of a system, it must be remembered that in most programs a fairly large number of individuals must be trained to support the system test program which can occur several years before system deployment. It is common practice for trainers/training devices to be designed and produced to support a recurring training program. Since a trainer is an end item in itself, it too requires establishment of a logistics support structure just like that for support equipment and the prime system. The training of operating and maintenance personnel can be seriously impeded if trainers are not usable because technical orders, spares, support equipment, facilities, trained operators, etc., are not available. The less than optimum training of system operators and maintainers could degrade the mission effectiveness and decrease system availability.

COMPUTER RESOURCES SUPPORT

The Computer Resources Support (CRS) element encompasses the facilities, hardware, software, documentation, manpower, and personnel needed to operate and support mission critical computer hardware/software systems. As both prime systems and support equipment increase in complexity, more and more software is being used. The expense associated with the design and maintenance of software programs is so high that we can't afford to not manage this process effectively. It is a general standard practice within each service to establish some form of a computer resource working group to accomplish the necessary planning and management of computer resources support. As can be seen in its definition, this element does cross the lines of responsibility in other ILS elements (i.e., facilities, manpower, etc.). It becomes a program office decision whether all the resource requirements needed to support this element are managed by a single CRS manager or by the other appropriate ILS element managers with the CRS manager monitoring.

FACILITIES

Facilities consists of the permanent and semi-permanent real property assets required to support a system, including studies to define types of facilities or facility improvements, location, space needs, environmental requirements, and equipment. Certainly the non-availability of facilities can be just as damaging to a system as would be the lack of spare parts, trained personnel, or support equipment. The main difficulties associated with this element are in funding and management responsibility. The process of facility design and construction is not routinely funded or managed by the program office but by civil engineering. This could range from base-level civil engineers for very small projects to the Army Corps of Engineers for projects of any significant size. Facility funds are, in the vast majority of cases, authorized by Congress for specific projects at specified location and are not transferable. The funding process takes 3 to 4 years to complete. Once the process has started, and certainly the closer the time comes to actual construction, the greater the impact will be from any change to facility location. A last-minute decision to deploy a system to a different locale may well require extraordinary DoD or even Congressional action to correct facilities delays. Keep in mind, facility requirements can range from the simple addition of 28 volts DC power to an existing work area, to the design and construction of a multimillion-dollar facility. In either case, the absence of the necessary capabilities within a facility or the absence of the facility itself is adversely felt by the prime system the facilities are intended to support.

PACKAGING, HANDLING, STORAGE, AND TRANSPORTATION

Packaging, handling, storage, and transportation (P,H,S,& T) is the combination of resources, processes, procedures, design considerations, and methods to ensure that all system, equipment, and support items are preserved, packaged, handled, and transported properly, including environmental considerations, equipment preservation for the short and long storage, and transportability. Packaging is more than cardboard boxes and styrofoam peanuts. Some items require special environmentally controlled, shock isolated containers for transport to and from a repair facility. A single package like this can cost tens of thousands of dollars. It also comes as no surprise that these types of reusable, repairable containers would also need spare parts, technical data, support equipment, etc., for their own support. P,H,S,& T may be a somewhat overlooked element, but it's not cheap. The reliability of a component can be significantly influenced by how its packaged, what type of handling equipment and procedures are used, where and how it is stored and the mode of transportation used to get it from the vendor to the eventual user. Transportability, on the other hand, means designing into a system or an item the ability to be transported. Trying to routinely transport an item, such as an intermediate maintenance avionics test station, that was not designed to be transported can result in the inoperability of the test station and therefore degradation of the repair capability for the items using the tester. Transportability requirement decisions must be made early in the system acquisition process and thoroughly delineated in the system specification.

DESIGN INTERFACE

The Design Interface is the relationship of logistics-related design parameters to readiness and support resource requirements. The logistics-related design parameters include reliability and maintainability, human factors, system safety, survivability and vulnerability, hazardous material management, standardization and interoperability, energy management, corrosion, nondestructive inspection, and transportability. These logistics-related design parameters are expressed in operational terms rather than inherent values and specifically relate to system readiness objectives and support costs of the system. Design interface really boils down to evaluating all facets of an acquisition from design to support and operational concepts for logistical impacts to the system itself and the logistic infrastructure.

This explanation is only a thumbnail sketch of the ILS elements and ILS management process. Each element has its own set of process, procedures, and techniques for use in satisfying the requirements of the individual element. However, one must not forget that the "I" in ILS stands for "Integrated." No program, and more specifically no logistics element manager, can afford to be so myopic in the management of the individual element(s) that the extensive interrelationship of the elements is forgotten. The administration of any of the logistics elements is a two-fold process: the individualized management and attainment of the element and the optimization of each element to the other elements applicable to a program. The Integrated Weapon System Master Plan (IWSMP) supports and documents the ILS process throughout the system life cycle.

It is common DoD practice to have total responsibility for a systems acquisition levied on a program manager or program director. These responsibilities are cost, schedule, performance, and supportability. It is also common for the program manager/director to delegate most, if not all, the responsibility for establishing the logistics support system to an integrated logistics support manager. This person usually has a moderate to intensive background in acquisition logistics and, depending upon the size of the acquisition program, has other logisticians under their control to perform all the tasks necessary to identify, plan, and implement the logistics support system.

14.6   Summary

The concepts, processes and procedures of ILS and LSA are certainly no panacea for all the ills associated with the acquisition and support of new or developing systems. The precepts of ILS do provide a process which can more thoroughly examine the requirements for supporting equipment vital to our defense needs. This thorough, systematic approach to logistics support is mandatory in the prevailing climate of cost-effectiveness.

References

[1] DoDI 5000.2, "Defense Acquisition Management Policies and Procedures."

[2] AFR 800-2, "Acquisition Program Management."

[3] AFR 57-1, "Air Force Mission Needs and Operational Requirements Process."

Chapter 15

MANUFACTURING MANAGEMENT

15.1 Introduction

Manufacturing has typically been an area that does not receive near the emphasis that it should in the acquisition process. Manufacturing, as with most of the other functional areas, needs to be considered early enough to make impacts on the design. It is not unusual to have an excellent design that is either too costly to produce or unproducible; things that we must strive to avoid.

It is interesting to note that manufacturing has been rolled under the umbrella of systems engineering in the new DoDI 5000.2, and in planning for the new Air Force Materiel Command. While this has a great many manufacturing individuals concerned that they will be even further buried, it also has the potential to help get manufacturing the early on recognition that it deserves.

One final observation about manufacturing problems, then on to the reading. One additional problem that we tend to have in manufacturing, producibility in particular, is that many of our engineers are given many classes in design techniques and strength of materials, but few have much background in manufacturing techniques. Manufacturing personnel many times have Industrial Technology degrees, learning much about manufacturing processes, but little about design and materials. Why is this important? It is clear that we need a combination of the two types of individuals, each with some knowledge of the other's field to make the process work effectively. Will this balance occur with the restructuring discussed above? Only time will tell.

15.2 Manufacturing Management Process

INTRODUCTION

Manufacturing management, also called production management, needs to be understood by most program personnel. The first time you address manufacturing issues should not be when your program starts gearing up for production. Manufacturing management is a process that can start as early as the concept exploration and definition phase and continue throughout the acquisition life cycle. This reading addresses two main areas: first, some of the most important manufacturing activities that occur during each phase of the acquisition process; and second, several manufacturing-related initiatives that affect the management of DoD acquisition programs. The material on manufacturing activities is from the DoD Manufacturing Management Handbook for Program Managers, published by Defense Systems Management College.

The term manufacturing, or production, has wide-ranging application anytime scarce resources are being transformed into goods and/or services. We will now take a look at the role of manufacturing management within DoD.

MANUFACTURING

For our purposes, manufacturing is the transformation of raw materials into products needed to perform DoD's national security mission. The resources required for this transformation process include people, equipment, money, material, facilities, manufacturing technology and processes, and time.

MANUFACTURING MANAGEMENT

Management of the manufacturing process is a subset of the larger function of program management and represents the techniques of economically planning, organizing, directing, and controlling the resources needed for production. A primary goal of manufacturing management is to assist program managers in assuring that defense contractors deliver goods and services of specified quality on time and within cost constraints agreed upon. To accomplish this goal, manufacturing managers must become involved early in the acquisition life-cycle of a program.

This early involvement is crucial if the DoD expects acquisition programs to function smoothly during the production phase where the largest portion of acquisition dollars is expended. Early involvement in the manufacturing function provides the opportunity to identify and reduce many of the major risks that have caused many DoD programs to falter during the

Table 15.1: Manufacturing life cycle activities.

CONCEPT EXPLORATION	DEMONSTRATION AND VALIDATION	ENGINEERING AND MANUFACTURING DEVELOPMENT	PRODUCTION
Evaluate production feasibility.	Reassess feasibility.		
Assess production risk.	Resolve production risk.		
Develop manufacturing strategy.	Preliminary manufacturing plan.	Finalize manufacturing plan.	
	Preliminary producibility plan.	Finalize producibility plan.	
		Conduct production readiness review(s).	Execute manufacturing program.
			Perform production surveillance.

transition from development to production, and experience substantial cost overruns, schedule delays, and performance compromises. By the Engineering and Manufacturing Development Phase, 85 percent of the acquisition decisions affecting major operation and support costs have already been made. This means that manufacturing issues need to be addressed as early as possible. Otherwise, changes in the latter part of EMD or early production will be very costly. Table 15.1 shows how manufacturing management affects the life cycle phases of the program. We will look at each phase individually.

15.3   Concept Exploration and Definition

EVALUATION OF PRODUCTION FEASIBILITY

One of the first activities performed in the concept exploration and definition phase is to evaluate production feasibility. The program manager ensures that a manufacturing feasibility assessment is accomplished in the initial phases of product development. The feasibility estimate determines the likelihood that a system design concept can be produced using existing manufacturing technology while simultaneously meeting quality, production rate, and cost requirements.

The feasibility analysis involves the evaluation of such factors as:

- Producibility of the potential design concepts.

- Critical manufacturing processes and required special tooling development.

- Test and demonstration required for new materials.

- Alternate design approaches within the individual concepts.

- Anticipated manufacturing risks and potential cost and schedule impacts.

The feasibility assessment is accomplished to relate the manufacturing risks incurred in selecting a particular design, fabrication concept, and material as the basis for moving into the demonstration and validation phase. Without this type of risk, the program manager may find that later phases of the program cannot be accomplished within the defined thresholds as a result of incompatibilities between the system design and the manufacturing technology available to execute it.

ASSESSMENT OF PRODUCTION RISKS

Based upon the feasibility assessment, the program management office develops a manufacturing risk evaluation to quantify the statement of manufacturing feasibility. Manufacturing risk assessment is a supporting tool for the contractor and program office decision making process. It seeks to estimate the probabilities of success or failure associated with the manufacturing alternatives available. These risk assessments may reflect alternative manufacturing approaches to a given design or may be part of the evaluation of design alternatives, each of which has an associated manufacturing approach. It should also consider the sensitivity of the feasibility estimates to the assumptions which were made on those areas of the design for which specific design data was not available.

The quantified risk levels (i.e. High, Medium, Low) can then serve as the basis for the development of specific risk resolution approaches for the later phases of the acquisition cycle and can provide guidance to the budget estimation process. Programs that have not addressed manufacturing risk during the development phases have had problems during the production phase. These problems can involve high cost, extensive design changes, unplanned material and process changes, and difficulties in delivering hardware on time which conforms to the contract requirements.

IDENTIFICATION OF MANUFACTURING TECHNOLOGY NEEDS

The evaluation of manufacturing capability is based on the analysis of the compatibility of the manufacturing facility and equipment with the demands of the manufacturing task. One result of the manufacturing feasibility evaluation is the identification of manufacturing technology needs, gaps between current capabilities and desired capabilities. Potential technologies existing today include examples such as:

- Thermoplastic-matrix composites.

- Composites out of autoclave bonding.

- Powder metallurgy.

- Graphite/epoxy composites.

- Aluminum metal matrix composites.

- Noncontact laser inspection.

- Carbon-carbon composites.

- Moldline fidelity technology.

Needs are identified so that the manufacturing capabilities that are required can be put on line in the factory prior to the production phase. When manufacturing technology development programs involve some risk, the program manager should consider requiring the design contractor to identify (or develop) fall-back positions for each of the risk areas and/or demonstrate the required capability in the laboratory or in pilot production.

DEVELOPING MANUFACTURING STRATEGY

This involves developing a manufacturing strategy for the system that is to be produced. Program manufacturing strategy is a subset of the overall acquisition strategy. Specific decisions are made concerning the level of competition to be attained during the production phase. If the program will be dual-sourced, the early planning must take into account whether capable alternate sources already exist or must be developed and actions needed to ensure that a suitable data package is available.

New manufacturing technologies, if required by the system concept, require specific plans for development, proof, and transition of the technology to the eventual producer. This effort necessitates close coordination with the service manufacturing technology organization to assure compatibility of the technology development schedule with the system development schedule. Studies have shown that competition is a major contributor in reducing weapons system cost. If competition is to be effective, it must result from the application of a clearly defined strategy to ensure that an environment of true competition is established and maintained.

15.4 Demonstration and Validation

REASSESSMENT OF PRODUCTION FEASIBILITY

Production feasibility is the likelihood that a system design concept can be produced using existing production technology while simultaneously meeting quality, production rate, and cost requirements. As a follow-on to the feasibility assessment accomplished during the concept exploration phase, the program office uses the increasingly more complete description of the system to update the assessment. This may be done within the program office or by the prime contractor. As the system design concept and manufacturing approach are validated and design decisions are made, the amount of flexibility on the choice of production technologies decreases. It is important for the program manager to ensure that design decisions reflect currently available production technology. Consideration of feasibility must occur in a bounded environment. The primary bounds are the existing state of production technology, the cost targets established for the system, and the production rate and schedule requirements.

Feasibility assessment is useful in supporting decisions concerning which of the competing system designs is carried into engineering and manufacturing development (EMD). It is also used to determine which of the manufacturing processes should be proofed during EMD and the nature of the proofing required. The process of weapons system design is dynamic and the search for the best solution often involves changes to the design concept that impact the manufacturing processes to be used. Failure to assess feasibility at a number of points during the acquisition process can result in accepting changes to the design which are incompatible with the capability of the industrial base.

ACCOMPLISHMENT OF PRODUCTION RISK RESOLUTION

Production risk resolution involves demonstrating the attainability of the levels of manufacturing capability required. During this phase, it is not necessary that all the details of the production processes be demonstrated. The areas that represent advances beyond the current capability should be demonstrated in environments which are somewhat representative of the production floor. The focus is on determining that there is a reasonable expectation that the manufacturing materials and processes required can be obtained or fabricated in sufficient quantity and quality to meet the production phase requirements. For instance, many raw materials are integral parts of our weapons systems, yet we import them, many exclusively, from many unreliable sources. These are important concerns when considering the production risk of an acquisition end item. Deferring risk resolution to a later phase incurs a concern that the design may go into production relying on the processes or materials which have relatively unpredictable processing time and cost. There is the possibility that compromising efforts to meet quality, cost, and schedule goals may adversely affect technical performance of the end item.

COMPLETION OF MANUFACTURING TECHNOLOGY DEVELOPMENTS

For those technologies identified during the concept exploration phase as requiring development, laboratory demonstrations are accomplished. Like the system development program, manufacturing technology development often requires a phased approach to definition and demonstration. The technology developer demonstrates that the required process or material capability is attainable under laboratory or controlled conditions and also describes the procedure by which the technology can be extended into the manufacturing shop environment. Since it is normally anticipated that critical processes are demonstrated in the production environment during the engineering and manufacturing development phase, it is important that the laboratory (or controlled production) process capability be demonstrated during this phase. Failure to do so may increase the risk, during EMD, that the material or process may be found not to be a viable approach for meeting the weapon system design requirements.

PRELIMINARY PRODUCIBILITY PLANNING

Producibility is a measure of the relative ease of producing a product or system. It is also an engineering function, often referred to as concurrent engineering or simultaneous engineering, directed toward generating a design which is compatible with the manufacturing capability of the defense industrial base. Each competing design needs to be evaluated from a producibility standpoint. The producibility effort must take into account the quantity of units or systems to be produced and the rate at which they will be manufactured, since quantity and rate determine the magnitude of the potential manufacturing efficiencies to be gained or problems to be avoided. Producibility reviews serve as a basis for estimating the likely manufacturing cost and assessing the level of manufacturing risk of the system. Results of these assessments support the development of specific contractual provisions for the EMD phase. Specific requirements may be identified based upon the inherent level of producibility, the specific system designs, and the susceptibility of each to manufacturing cost reduction through an aggressive producibility program.

DEVELOPMENT OF PRODUCTION READINESS REVIEW PLAN

One of the major program office tasks during the EMD phase is the production readiness review (PRR), in accordance with DoDI 5000.38. It is critical that the specific requirements for contractor planning and support to the PRR are included in the EMD contract. There is also a need to ensure that the necessary Government evaluation skills are available during EMD. These needs can only be met if the major readiness issues are identified during the demonstration and validation phase and the methods for evaluating readiness are clearly defined. The readiness issues must cover both the defense system design and the production planning required. Since many of these issues are normally evaluated as part of the continuing process of design and program reviews, the planning for PRR should clearly describe how the outputs and analyses of these reviews can be applied to the PRR tasks.

15.5   Engineering and Manufacturing Development

MANUFACTURING PLAN

At the end of EMD, all of the information necessary to plan the detailed manufacturing operations for the system should be available. This information is described in a manufacturing plan covering the issues of manufacturing organization, make or buy planning, subcontract management, resources and manufacturing capability, and the detailed fabrication and assembly planning. The plan also describes the types of Government furnished property (GFP) required and the specific need dates for it. The contractor management control system plan, including those for configuration management, the control of subcontractors, and for manufacturing performance evaluation is described in sufficient detail for the program management office to determine their expected utility. The plan also includes consideration of the potential requirements for industrial preparedness, including surge capability during the production phase, and the post-production phase requirements for support to employment of the system in combat situations. The development of this formal manufacturing plan contributes value to the program from two standpoints. The primary benefit accrues from the fact that the contractor has to crystallize the manufacturing planning to a point where it can be described in the detail required. The secondary benefit is the usability the plan provides to the program management office personnel; it serves as a basis for a structured review of the contractor approach, the expected cost of the production phase effort, and a full assessment of manufacturing risk. Where such a plan is not developed during the EMD phase, there is often unnecessarily high cost and schedule turbulence at the front end of the production phase.

PRODUCIBILITY ENGINEERING AND PRODUCTION PLANNING

Producibility, as noted previously, is a measure of the relative ease of producing a product or system. Alternate manufacturing methods, materials, resources, and processes must be a consideration of the detailed design if the economics of manufacturing and assembly are to be considered. Producibility studies and analysis of the alternatives are conducted by the contractor with consideration of the impact on cost schedule and technical performance. Early production planning based on design and schedule requirements is essential if production delivery schedules are to be fulfilled. Production planning must include identification of potential problems with an assessment of the capability required to produce the item and industry's current capability to manufacture the system as designed. Potential production problems that require further resolution by study or development must be identified and action for resolution initiated. The producibility engineering and planning effort also results in the definition and design of the special tooling and test equipment required to execute the production phase effort as well as the preparation and release of the manufacturing data package required for the start of manufacture.

There are a number of factors to be considered in ensuring the producibility of a design:

- Liberal tolerances (dimensions, mechanical, electrical).

- Use of materials that provide optimum machinability, formability, and weldability.

- Shapes and forms designed for castings, stampings, extrusions, etc., that provide maximum economy.

- Inspection and test requirements that are the minimum needed to assure desired quality and maximum usage of available and standard inspection equipment.

- Assembly by efficient, economical methods and procedures. Minimized requirements for complex or expensive manufacturing tooling or special skills.

There should be evidence that the contractor has accomplished producibility analyses of various options for the manufacturing task. The EMD phase results in the system design for entering production. As the design evolves during EMD, its producibility is subjected to regular review (as part of the normal design review process).

DETAILED PRODUCTION DESIGN

Prior to release of drawings to the manufacturing location, the detailed design drawings, bills of material, and product and process specifications must be completed. Further, it is essential that design reviews be conducted to assure that the contractor is complying with the design requirements and meeting the cost/design goals. The final design definition is the result of the performance requirements, the outcomes of the testing accomplished, producibility studies, and other design influences. The production phase effort requires that the design be specified in sufficient detail so that the required processes and resources can be identified and obtained.

PRODUCTION READINESS REVIEW

One of the ways the Government assesses production planning efforts is to accomplish production readiness reviews (PRR). The objective of a Production Readiness Review is to verify that the production design planning and associated preparations for a system have progressed to the point where a production commitment can be made without incurring unacceptable risks of breaching thresholds of schedule, performance, cost, or other established criteria (DoDI 5000.2). PRRs are conducted by a government team as a time-phased effort that spans engineering and manufacturing development and encompasses the developer/producer and major subsystem suppliers. The PRR examines the contractor's design from the standpoint of completeness and producibility. It examines the producer's production planning documentation, existing and planned facilities, tooling and test equipment, manufacturing methods and controls, material and manpower resources, production engineering, quality control and assurance provisions, production management organization, and controls over major subcontractors. The results of the PRR assist the program manager in assessing the risk involved in transitioning into production, i.e., that the system is ready for efficient and economical rate production.

15.6   Production and Deployment

MANUFACTURING PROGRAM

The primary function of the production phase is to complete the manufacture of the defense system within the established time and cost constraints. Normally, the production rate is structured to start slowly (i.e., Low-Rate-Initial-Production (LRIP)) and build into a defined steady state rate. Much of the same type of evaluation of contractor planning for initiation of the production phase (generally through the PRR) needs to be focused on the contractor planning to increase to the defined rate. The program manager also focuses attention on the levels of engineering change activity. An excessive number of engineering changes can disrupt the structure of the manufacturing planning and result in high manufacturing costs. Also, attention needs to be given to ensuring that acceptance criteria for the product or system are clearly specified and that there is minimum use of waivers, deviations, and Material Review Board actions during the acceptance process. The program office manufacturing personnel participate in the physical configuration audit (PCA) when the "as built" item is compared with the technical documentation. Upon satisfactory completion of the PCA, the primary acceptance criteria is the physical and test requirements listed in the technical documentation.

The completion of the production phase normally involves a series of contract actions which are planned and completed to fill the system acquisition objective. For each of these contracts, a decision is made on the contract type, the incentive structure, if any, the level of Government control, and the desired program visibility.

PRODUCTION SURVEILLANCE

One of the primary program management tasks during this phase is to establish and maintain a system for accomplishing surveillance over the progress of the contractor performing the manufacturing tasks. Generally, the program manager wants to ensure that information is available to measure contractor effectiveness from time, cost, and technical achievement standpoints. The program manager must also choose between a formally structured and contractually specified management control system or a currently existing contractor system. When problems occur during the production phase, the management control system should provide timely information to the program manager in a format that supports decision making and action processes.

PRODUCT IMPROVEMENT

One of the methods used to support the decision making process is follow-on operational test and evaluation (FOT&E). The FOT&E and the initial user feedback on the system often identify areas where improvements can be made to the system to allow it to better meet the constantly changing operational environment. The challenges for the program manager involve the decisions on which of these improvements to make, and the method of incorporating them on the production line. To minimize production cost, the number of engineering changes should be kept to a minimum but operational requirements often favor change.

A program may also involve preplanned product improvement (PPI). If this acquisition strategy applies, when and how to incorporate such improvements must be resolved early in the program.

15.7   Management Initiatives

We will now look at several manufacturing initiatives that are affecting the management of DoD acquisition programs.

INDUSTRIAL MODERNIZATION INCENTIVE PROGRAM

The Industrial Modernization Incentive Program (IMIP) received its initial impetus during the late 1970s. At that time there was significant national concern over the declining industrial base, especially in the defense sector. Classical forces of competition were insufficient to stimulate productivity-related investments. Further, DoD business arrangements did not always provide adequate incentives for productivity investments. In a search for ways to motivate contractors to improve their productivity, the Air Force developed "Tech Mod" (or technology modernization), and the Army developed what was called "industrial productivity initiatives." In 1982, the term IMIP was coined in an attempt to combine these endeavors into an integrated, DoD-level effort. A "test period" began during which waivers to acquisition regulations were considered, new and unique ways of contracting were developed, and new methods were established to derive incentives and share the savings with contractors. By 1985, DoD Directive 5000.44 and a DoD guide for the implementation of IMIP were under development. These documents established the Industrial Modernization Incentive Program (IMIP) as a DoD program to systematically implement new technologies into the defense industry.

The definition of IMIP has two important characteristics. First, IMIP is a joint venture, a cooperative effort between Government and industry. Second, IMIP involves the negotiation of a business deal between partners whereby each partner contributes funds and assumes risks, but also shares benefits in a "win - win" fashion.

IMIP has short term as well as long term objectives. Short term objectives include increased product quality, reduced lead times and lower costs for weapons systems. The long term objective is a strong, healthy industrial base that can meet surge and mobilization requirements if a conflict or war should arise.

IMIP PHASES

IMIP programs are accomplished in three phases. Phase I begins with a "Top down factory analysis." This analysis evaluates the needs of the overall facility, locates bottlenecks, and identifies potential manufacturing technologies and modernization opportunities which apply to the type of work normally done at that factory.

Phase II develops the enabling technologies and work centers that are needed to take advantage of the opportunities identified during Phase I. Prototypes are built and their use is demonstrated during a formal review attended by the Government. Plans for implementing the new technologies onto the production floor are also accomplished.

Phase III implements the IMIP through contractor purchase and installation of capital equipment in the factory. Actual benefits are monitored and as they accrue, incentives flow to the contractor, savings flow to the Government, and the new technology is made available for transfer to other interested defense contractors.

Funding of these endeavors is shared between the Government and industry. For example, during fiscal years 1981 - 1987, $1.9 billion was invested within the Aeronautical Systems Division alone. Of that amount, private industry invested about $1.5 billion, or roughly four times that which the Government invested. Validated savings thus far exceed a quarter of a billion dollars. However, most IMIP projects have not yet entered

Phase III. The projected savings over the next decade is about $5 billion, or over two and a half times the original investment. While our experience is still somewhat limited, documented results of this program appear to promise a strong, positive impact on the industrial base.

PRIDE PROGRAM

In 1990, Air Force Systems Command created a new program called Program to Revitalize Industrial Development and Efficiency (PRIDE) which is basically an outgrowth of IMP. The PRIDE is designed to motivate subcontractors to develop critical manufacturing technologies in support of major defense contractors. To date only a limited amount of data is available on this new program but initial results are very positive.

WORK MEASUREMENT

In the early 1970s, MIL-STD-1567A was developed to provide a framework for disciplined work measurement systems on defense contracts. Faced with increasing costs and decreasing buying power, DoD hoped to take the guesswork out of estimating costs and to actually reduce costs. The following discussion highlights the essence of what work measurement is all about.

First, work measurement is a method for evaluating efficiency that determines the amount of time it "should take" to do a given job. This "should take" time is expressed by what is commonly known as a labor time standard. Next, the actual time needed to do the job is measured and compared to the standard. This comparison is normally expressed one of two ways. If standard hours are divided by the actual hours taken to do the job, the result is an efficiency rating. Within the defense industry, you are likely to hear reference to the other type of comparison. This method simply inverts the equation and calls the result a performance index or realization factor.

Our definition of work measurement is not complete until we have analyzed the results of the comparison between standard and actual performance. Whenever efficiency is low or the performance index is high, we have what is known as variance. Since this information is collected at various levels in the work breakdown structure, it is possible to pinpoint areas in the factory where problems and inefficiencies exist. Correction of these problems may involve training of workers, method improvements, revised layout of the facility, changes in management procedures, or whatever else the source of the problem turns out to be.

As mentioned, a labor standard defines the amount of time that a given task "should take." However, there are additional requirements that must be met if we are to have a standard that can be applied accurately and consistently. The following definition fully describes all the characteristics of a valid labor time standard:

> A labor time standard defines the time it should take a normally skilled operator, following a prescribed method working at a normal all-day level of effort, to complete a defined task with acceptable quality.

Note that the phrase "all-day level of effort" implies something that we have not mentioned yet. A standard time is composed of the sum of what is called the "normal time" needed to do a job and some additional time for "allowances." These allowance recognize three factors that slow a worker down: (1) the need for personal time to use the restroom, take a coffee break, etc.; (2) fatigue that occurs as the day wears on; and (3) minor, unavoidable delays that are not under the worker's control. Therefore, normal time plus some percentage for allowance constitute a standard time.

WORK MEASUREMENT STANDARDS

MIL-STD-1567A, which is the contractual document for implementing work measurement, recognizes two types of time standards, referred to as type I and type II. Type I standards are derived through a number of recognized industrial engineering techniques such as time study, predetermined time systems, standard data, and work sampling. Each technique has specific strengths and applications.

Type II standards are estimates based on experience and historical projections. While less reliable, type II standards are effective during the early phases of a program and when it would not be cost effective to develop a type I standard for a complex, low-volume task.

BENEFITS

DoD hopes to gain a number of benefits from work measurement programs. By knowing more accurately just exactly how long a task should take, it is possible to:

- Improve cost estimates.

- Improve the accuracy and credibility of DoD budget requests.

- Improve scheduling and other manufacturing control activities.

- Help in the solution of layout and material handling problems by providing accurate figures for utilization of equipment.

Through "variance analysis," or the performance analysis referred to earlier, it is possible to:

- Reduce costs.

- Improve manufacturing methods and processes.

- Reduce scrap and waste.

- Evaluate worker performance and establish wage incentives.

A final, but important, question to ask is, "Do work measurement programs pay for themselves?" Private industry has resisted the implementation of work measurement on DoD programs, in part because they are afraid that the system will grow into a costly, burdensome reporting nightmare. The Government takes a more positive view, estimating that the savings potential ranges from 10 to 20 percent of direct labor cost, while the cost of work measurement programs is about 1 to 4 percent of direct labor. We are still gaining experience in this area, but industry surveys suggest savings to cost ratios of from 2:1 to 5:1. Only time will reveal whether industry fears of a costly and burdensome reporting system are well founded.

TRANSITION FROM DEVELOPMENT TO PRODUCTION

Our final subject addresses an area that is highly likely to become "institutionalized" as a way of doing business by all three military services. In 1982, Mr. Will Willoughby was selected as chairman of a defense science board task force on the subject of transition from development to production. We have the results of the study from that task force available to us in DoD manual 4245.7, dated September 1985.

Two things about the approach taken by the task force are particularly appealing. First, the results are practical. In the words of Mr. Willoughby, "Many acquisition studies have gone 'cosmic,' that is, they have made suggestions that are beyond the ability of program managers, or even the DoD itself, to control." This study offers specific tools that managers can employ within the existing system. We do not have to rely on the unlikely possibility that someone else will change the system for us.

Second, this approach integrates the entire acquisition process rather than sub-optimizing on only one aspect of it. The remainder of this section describes the problems that needed to be solved and the tools that the task force provided to do so.

An often discussed aspect of the acquisition process is the length of time it takes to develop and deploy weapons systems. Although there have been numerous attempts to shorten this cycle, if anything, it has only grown longer. The reasons for shortening the cycle are directed mainly toward cost, and to some extent, though not enough, toward readiness. Although the long acquisition cycle is certainly not desirable, it might be tolerable if the process yielded satisfactory results. But many new weapons systems do not perform as "originally advertised," and often require burdensome maintenance and logistics efforts.

Often, the first evidence of weapon system problems does not become apparent until the program attempts to transition from engineering and manufacturing development into production. Most acquisition managers seem to recognize that there is a risk associated with the transition, but may not know the magnitude nor the origin. The task force felt that some of this uncertainty occurs because the transition is not a discrete event, but a process composed of three elements: design, test, and production. Early involvement of the manufacturing process provides the opportunity to reduce the risks associated with this transition. Many programs simply cannot succeed in production despite having passed the required milestone reviews. A poorly designed product cannot be efficiently tested, produced, or deployed. In the test program there are far more failures than should be expected. Manufacturing problems overwhelm production schedules and costs. The best evidence of this is the "hidden factory syndrome" with its needlessly high redesign and rework costs.

Corrective measures by DoD have focused on various management checkpoints and review activities. Gradually, numerous layers of management have been added that have tended to compartmentalize and polarize the major areas of the acquisition process, namely: design, test, and production. The task force concluded that the causes of acquisition risk are more technical than managerial. The key to reducing risk is in "disciplined engineering" throughout these interrelated processes of design, test, and production. Failure to do well in one area results in failure to do well in the other areas.

CRITICAL EVENTS

The task force, represented by DoD as well as private industry, drew upon their combined experience to generate a matrix of critical events in the design, test, and production processes. These events were then transformed into templates. A template can be expanded to describe three things for each event: (1) areas of risk, (2) an outline to reduce the risk, (3) a time line that shows when the activity should occur during the acquisition cycle. Risk is introduced to the program whenever a particular template activity begins late or does not finish on time. The major areas of funding, design, test, production, facilities, logistics, and management were chosen through analysis of recurring problems on a substantial number of programs. Funding is presented first because it influences every other template in the transition document. There is a wealth of information and experience reflected in the templates. The manual that documents this information, DoD 4245-7, should be a key reference available to all acquisition managers.

References

[1] NAVSO P-6071, "Best Practices: How to Avoid Surprises in the World's Most Complicated Technical Process," March, 1986.

[2] DoDI 5000.2, "Defense Acquisition Management Policies and Procedures."

[3] DoD "Manufacturing Management Handbook for Program Managers," Defense Systems Management College, July 1984.

Chapter 16

TEST AND EVALUATION

Test and Evaluation (T&E) helps reduce acquisition program risk and is one of the most important areas of the acquisition process. Thorough planning is required in order to collect quality information for analyses and evaluations and to keep senior decision makers fully informed. Test and Evaluation gives you that information. Recent acquisition reform legislation, like the Packard Commission Report, emphasizes the importance of testing the system as early as possible before locking into concrete decisions. This chapter addresses the concept of testing in the acquisition environment. The purpose of Development and Operational Test and Evaluation is introduced, as well as how these activities fit into the acquisition life cycle. Next, the role of one of the operational implementers of test policy, the Air Force Operational Test and Evaluation Center (AFOTEC), and the documentation that is important in providing test plans and results, are addressed.

We must be concerned about systems requirements as well as test and evaluation of systems because they help reduce acquisition program risk and ensure that more operationally effective and suitable systems are fielded. Numerous subjective judgements go into validating whether or not the threat simulators and targets really simulate hostile forces, and whether the tests are realistic enough, despite the artificiality of the test rules of engagement.

16.1   Test and Evaluation Process

Test and Evaluation (T&E) begins as early as possible in the acquisition process and continues throughout the entire system life cycle. According to DoDI 5000.2, test planning must begin during Phase), Concept Exploration. Moreover, sufficient T&E must be accomplished successfully before decisions are made to commit significant additional resources to a program or to advance it from one acquisition phase to another. While conducting T&E, quantitative data is used to the maximum extent possible, thereby minimizing subjective judgments. What are the main purposes of T&E? Essentially, it encompasses:

1. The assessment and reduction of risks;

2. The evaluation of the system's operational effectiveness and suitability; and

3. The identification of system deficiencies.

As you can see, that is not only a mouthful to say, but also a Goliath to conquer. Nonetheless, the difficulty of the task is probably equal to its importance to the program. It allows early evaluation of the program's technical feasibility and operational utility, and it facilitates earlier, less costly correction of system deficiencies.

Additionally, at the Defense Acquisition Board (DAB)/Air Force Systems Acquisition Review Council (AFSARC) reviews, much of the information in the Integrated Program Summary (IPS) used to assess program progress is derived directly from T&E results. The bottom line, then, is that the results from T&E significantly affect the future of the program—whether it advances, changes, or dies.

Test and Evaluation results are the primary source of information Air Force Program Managers and DoD decision makers use to make important decisions. For instance, the results of Test and Evaluation can greatly enhance program evaluations concerning cost, schedule, technical performance, and logistic supportability. Thus, the primary purpose of all Test and Evaluation is to make a direct contribution to the timely development, production, and fielding of systems that meet the user's requirements and are operationally effective and suitable.

STRUCTURE

Within a SPO, a typical T&E directorate is not a simple thing to define. Depending on the number of programs involved and their size and complexity, the organization can vary from a deputy director for large programs, a chief of a division for other small programs, or possibly only one or two individuals for small or one-of-a-kind programs. In any case, while the complexity, schedules, and resource planning may change, the mission of the organization usually does not. Regardless of organization type, the "testers" must plan, coordinate, and manage the program test activities within policies set by the DoD and service headquarters. The larger programs usually require more schedule and test disciplines due to more test articles and, possibly, more complex operations. However, testing of smaller programs should receive the same emphasis within the PO as the large programs.

TYPES OF TESTING

There are essentially two kinds of T&E: Development Test and Evaluation (DT&E) and Operational Test and Evaluation (OT&E). OT&E associated with an research and development program is further sub-divided into two phases: Initial Operational Test and Evaluation (IOT&E) and Follow-on Operational Test and Evaluation (FOT&E). The transition point for these two phases is Milestone III (Production decision). Finally, for certain programs/subsystems such as non-research and development funded programs, joint programs, and computer software subsystems, modified or specialized testing requirements apply. We will now look at each of these T&E categories.

16.2 Development Test and Evaluation

Through Development Test and Evaluation, the Air Force demonstrates that system engineering design and development is complete, design risks have been minimized, and the system performs as required and specified. DT&E includes the test and evaluation of components, subsystems, hardware/software integration, associated software, and pre-production models of the systems. It involves an engineering analysis of the system's performance, including its limitations and safe operating parameters. Furthermore, the system is evaluated against engineering and performance criteria specified by the implementing command. Also addressed are the logistics engineering aspects (supportability, maintainability, reliability, etc.) as well as compatibility and interoperability with existing or planned systems.

DT&E usually starts in the concept exploration/definition phase and continues through the engineering and manufacturing development phase. DT&E may (and usually does) continue throughout the life cycle of a program, although more DT&E activity occurs early in a program. Furthermore, it may embrace testing of system improvements or modifications designed to correct deficiencies or to reduce life-cycle costs. According to DoDI 5000.2, the overall DT&E objectives encompass the following:

1. Identify potential operational and technological limitations of the alternative concepts and design options being pursued,

2. Support the identification of cost-performance trade- offs,

3. Support the identification and description of design risks,

4. Substantiate that contract technical performance and manufacturing process requirements have been achieved, and

5. Support the decision to certify the system ready for operational test and evaluation.

16.3 Operational Test and Evaluation

Operational Test and Evaluation is the T&E conducted to determine a system's operational effectiveness and suitability under realistic combat conditions and to determine if the minimum operational performance requirements as specified in the Operational Requirements Document have been satisfied. Operational Test and Evaluation is conducted throughout the system life cycle in as realistic conditions as possible. OT&E uses personnel with the same types of skills and qualifications as those who will operate, maintain, and support the deployed system. As previously mentioned, OT&E includes two sub-phases: IOT&E and FOT&E.

INITIAL OPERATIONAL TEST AND EVALUATION

Initial Operational Test and Evaluation (IOT&E) is conducted to provide a valid estimate of a system's operational effectiveness and suitability and is normally completed prior to the first major production decision using Low-rate Initial Production (LRIP) articles. In the Air Force, AFOTEC is responsible for all IOT&E.

FOLLOW-ON OPERATIONAL TEST AND EVALUATION

Follow-on Operational Test and Evaluation (FOT&E) is usually conducted after the first major production decision or after the first production article has been accepted. It may continue throughout the weapon system's life cycle. AFOTEC is responsible for FOT&E of "selected systems," usually the large or special interest programs. The using major commands perform FOT&E on all other systems.

OPERATIONAL EFFECTIVENESS

In order to evaluate the operational effectiveness of a system, we must first understand what the term means. Operational Effectiveness denotes the overall degree of mission accomplishment of a system when it is used by representative personnel in the context of the organization, doctrine, tactics, threats, and environment in the planned employment of the system. Basically, we take a production representative system and check to see if it accomplishes its mission as planned, trying to make the operational environment as realistic as possible.

OPERATIONAL SUITABILITY

Operational Suitability is the degree to which a system can be satisfactorily paced in field use, with consideration being given to availability, compatibility, transportability, interoperability, wartime usage rates, maintainability, safety, human factors, manpower sustainability, logistics supportability, and training requirements. Although this seems close to the definition of effectiveness, the two terms have distinct differences.

Let us say, for example, that we have an aircraft that can deliver weapons to target with the desired accuracy and is able to avoid the threats in its environment. We could then say that this aircraft is operationally effective. Now, let us say that this same aircraft requires 100 hours of maintenance for each hour in the air, and that it requires special equipment to be able to start the aircraft. In this case, we would probably say that the aircraft is not operationally suitable. From this example you will hopefully be able to see the differences between the terms and why both are important.

16.4   Other Types of Testing

While development test and evaluation and operational test and evaluation are the two major types of testing on major acquisition programs, there are several other types of testing that you should be familiar with.

QUALIFICATION TEST AND EVALUATION

On programs where there is no research, development, test and evaluation (RDT&E) funding, Qualification Test and Evaluation (QT&E) is performed in lieu of DT&E, and Qualification Operational Test and Evaluation (QOT&E) is performed in lieu of OT&E. These programs might include Class IV/V modifications, simulators, software programs, and off-the-shelf equipment. QT&E is usually performed by the implementing command, while QOT&E is performed by the using command. * Essentially, the same test policies for DT&E and OT&E apply to QT&E and QOT&E.

QUALIFICATION TESTING

Qualification Testing is a subset of DT&E testing and is not related to QT&E. It verifies design integrity on pre-production items and verifies manufacturing process integrity on production items. Qualification testing is

---

*In the Air Force, all QOT&E is performed by AFOTEC

also the type of testing we perform on each configuration item to ensure that they meet their requirements.

PRODUCTION ACCEPTANCE TEST AND EVALUATION

Production Acceptance Test and Evaluation (PAT&E) assures that production items demonstrate the fulfillment of the requirements and specifications of the procuring contract or agreement. The testing also ensures the system being produced demonstrates the same performance as the preproduction models and operates in accordance with specifications. PAT&E is usually conducted by the Quality Assurance section of the program office at the contractor's plant and may involve operational users.

LIVE FIRE TEST AND EVALUATION

The requirement for Live Fire Test and Evaluation (LFT&E) is established by law which certain DoD weapon systems must satisfy before they can be produced in quantity and fielded. A live fire test is a test event so identified within an overall LFT&E strategy which involves the firing of actual munitions at target components, subassemblies, subsystems, and/or subscale or fullscale targets to examine vulnerability, and/or lethality issues including effects on both material and personnel.* The intent of LFT&E is twofold: (1) to assure that battle damage tolerance and damage control of our crew-carrying combat systems to actual threat weapons is known and acceptable; and (2) to assure that the lethality of our conventional weapons against actual threat systems is known and acceptable.

COMBINED DT&E/IOT&E

Combined DT&E/IOT&E is used when separate testing would cause significant delays or increases in systems acquisition costs and test resources. In this testing, the single article has a combined data base, but independent reports and analysis, and generally independent test objectives are being accomplished. This type of testing also requires a lot of advance planning to ensure that the proper types of information are available, and that the results will be accurate and unbiased. By law, separate, dedicated

---

*Vulnerability is the characteristics of a system that cause it to suffer a definite degradation (loss or reduction of capability to perform its designated mission(s)) as a result of being subjected to a certain level of effects in a manmade hostile environment. Lethality is defined as the ability of munitions to cause damage that will cause the loss of, or degradation in the ability of, a target system to complete its designated mission(s). A related concept of survivability is also used, which is defined as the capability of a system to avoid or withstand a manmade hostile environment without suffering an abortive impairment of its ability to accomplish its designated mission.

operational test events must be conducted to further evaluate the system's operational effectiveness and operational suitability.

16.5   Testing Organizations

There are a number of organizations involved in testing a weapon system, with various responsibilities residing with different groups in different testing phases. In this section, we discuss the various organizations involved in test and evaluation.

CONTRACTOR

In addition to the governmental agencies, the contractor plays a key role in the DT&E, especially in the early part of the test program. A contractual system test plan is developed jointly by the PO and the contractor and it identifies the roles of each participant. The contractor will be involved in sub-system testing, operational mock-up testing, certain key test events' such as the test that verifies a missile performs all of its deployment activities in the correct timing and sequencing, and a number of other tests leading up to the first live launch or first flight.

IMPLEMENTING COMMAND

The implementing command, which for the Air Force is normally AFMC, manages DT&E. AFMC assigns this responsibility to the Program Manager. The PM then designates a DT&E test director (sometimes the PM) to exercise control over the DT&E test team and associated resources. If the Program Manager needs another organization to perform testing, a Responsible Test Organization (RTO) is appointed by HQ AFMC.

RESPONSIBLE TEST ORGANIZATION

The Responsible Test Organization is responsible for the development test and evaluation of the system during engineering and manufacturing development, and prior to initial operational test and evaluation. The RTO is normally delineated in the Program Management Directive (PMD). The Program Manager or RTO may also receive assistance from a participating test organization (PTO), which is a secondary testing center that has capabilities the RTO doesn't. A good example of this is testing stealth technology items. Most of the flight testing will probably be accomplished at places like Edwards AFB, but testing of stealth characteristics is usually accomplished at a facility at Holloman AFB.

The RTO and participating test organization (PTO) are normally chosen from AFMC test resources, such as the Air Force Flight Test Center (AFFTC), Arnold Engineering and Development Center (AEDC), Air Force Development Test Center (AFDTC), 4950th Test Wing, and other laboratories and centers.

The RTO normally appoints a test manager and prepares a test plan. There may be a feeling at times, that the RTO has usurped the power of the Program Manager in its actions. The truth is that the RTO's function is to test and maintain a capability in test management and test facilities. The RTO test plan is submitted to the Program Manager for approval and then to the Air Force Operational Test and Evaluation Center (AFOTEC) for review and comment. The RTO then conducts the test with participation by AFOTEC and the user/support commands as required. The report is normally prepared by the RTO and may be approved for publication by the RTO or the Program Manager. Normal coordination procedures would seem to dictate that the Program Manager would have control over these reports. However, the centers are justly proud of their objective test capability and ability to publish their findings with integrity. It is up to the Program Manager to resolve any problems that impact on his or her program as a result of the RTO's rigorous tests.

AIR FORCE OPERATIONAL TEST AND EVALUATION CENTER

The Air Force Operational Test and Evaluation Center (AFOTEC) is a Field Operating Agency (FOA) headquartered at Kirtland AFB, New Mexico, that evaluates the operational effectiveness and suitability of weapon systems. Some important points about AFOTEC are the fact that it is an independent source of information separate from the developing or using command, and it reports directly to the Air Force Chief of Staff. The principle elements of AFOTEC are the headquarters and field test teams. Emphasis is placed on testing the weapons systems with objectivity and realism. This means that AFOTEC's role in the weapons system acquisition process is quite involved and important. AFOTEC conducts all IOT&Es and QOT&Es of all systems. AFOTEC may assist using commands in conducting FOT&E. On systems upgrades that are of special interest to the Air Force and OSD, AFOTEC may be tasked to accomplish the FOT&E.

USING COMMAND

The using command participates in both DT&E and OT&E by providing resources to help conduct the test program, managing QOT&E on system upgrade programs, and conducting follow-on operational test an evaluation throughout the life of the system to ensure that the system remains operationally effective and suitable.

TEST PLANNING WORKING GROUP

The Test Planning Working Group (TPWG) is chartered by the program office to coordinate testing between all agencies involved in or affected by the testing. One of its primary tasks is to write and update the Test and Evaluation Master Plan that contains the overall test strategy for the system. The contractor also becomes involved with the TPWG to help coordinate all test efforts.

16.6 Test and Evaluation Master Plan

The Test and Evaluation Master Plan (TEMP) is the primary document used to assess adequacy of planned DT&E. It contains a mission and/or system description, the required operational/technical characteristics, critical technical/operational issues, test management responsibilities, DT&E and OT&E outlines including objectives, and a test resource summary.

The PM prepares the TEMP with assistance from the Test Planning Working Group (TPWG). This group consists of personnel from the PO, test agencies, AFOTEC, operating and supporting MAJCOMS, contractor (when appropriate), and other involved test agencies. Besides providing assistance in the initial TEMP preparation, the TPWG meets periodically to update test plans and monitor test progress. The TEMP is coordinated with all affected agencies. For major systems the TEMP is approved at the Under Secretary of Defense level.

The Test and Evaluation Master Plan is a statutory requirement for alll acquisition programs and is one of the stand alone documents that is presented as part of the Integrated Program Summary (IPS) from Milestone I on.

16.7 Program Test Plans, Procedures, and Reports

Program test plans, procedures, and reports are primarily a contractor/ Government agreement on the test philosophy, the specific actions to carry out that test philosophy, and the reports showing the results of carrying out those test procedures. PM's are ultimately responsible for the completeness and accuracy of test plans, procedures, and reports put on contract and approved by the Government.

References

[1] DoDI 5000.2, Defense Acquisition Management Policies and Procedures.

[2] AFR 80-14, Test and Evaluation, 3 November 1986.

[3] AFP 55-43, Management of Operational Test and Evaluation, 28 June 1985.

[4] AFSCP 80-27, Summary of AFSC Major Ranges and Test Facilities, 26 January 1981.

Chapter 17

MAINTENANCE MANAGEMENT

17.1 Introduction

Equipment maintenance is the largest facet of logistics in terms of money, manpower, facilities, or any other resource one might consider. Maintenance is now recognized as a major function in unit capability because that capability is a function of both equipment readiness and mission ability. Without an effective maintenance operation, mission equipment can not be ready.

The purpose of this chapter is to understand more about this primary function of logistics, equipment maintenance. This section will review some important definitions. These definitions will be followed by an historical review of how the structure and concept of maintenance have evolved. Next, the three major functions of maintenance and the three classical maintenance tasks will be described. A description of the depot-level maintenance functions of the Air Force Materiel Command will follow. The chapter will conclude with a discussion of maintenance decision impacts.

According to Air Force Manual AFM 2-15, maintenance is the process of keeping materiel in a serviceable condition or restoring it to that condition when it fails or malfunctions. There are three primary maintenance tasks involved in this process – prevention, correction, and modification. Some of the specific maintenance tasks that are performed include servicing, inspection, calibration, testing, repair, overhaul, modification, and reclamation. Examples of these maintenance tasks will be covered later in this chapter when the three functions of maintenance are discussed.

The word *materiel*, as used in this definition, will be used to describe the following: (1) WEAPON SYSTEMS, such as aircraft, missiles, support equipment, munitions, ground communications, trainers, simulators, and tactical and special purpose vehicles, as well as (2) MAJOR EQUIPMENT ITEMS, such as instruments, life support, airborne electronics, meteorological equipment, fluid driven accessories (hydraulics), landing gear, and canopies. Therefore, the term "materiel" can be used to describe almost any major weapon system or equipment item. Another important word or concept in this definition is "process."

A process is something that takes inputs and converts them to some kind of output. The maintenance process requires facilities/shops, equipment/tools, specialized skills, procedures, materials, and information. The terms "facilities" and "shops" refer to the physical location where the maintenance process takes place. These can be as simple as a small repair shop or as extensive as depots. The equipment and tools are the "implements" that are used to perform the maintenance. As with facilities, these implements may range from the simple to the complex. They may be as simple as a screwdriver, or as complex as the F-16 Automatic Test Station. Any safety devices, such as protective eye wear, would also be included here.

The specialized skills represent the talents, skills, and training that each person who performs the maintenance either brings to the process or must be trained in before he or she can perform any maintenance actions. When discussing the maintenance process, the "procedures" inputs represent the instructions, the guidance and guidelines, the step-by-step procedures, and all of the documentation that is necessary for the maintenance process to function. These can be common practice procedures, such as unplugging electrical items before performing any work, or sophisticated engineering specifications that would be required by the depot before it could manufacture a replacement part.

The "materials" input, as used here, has a different meaning from the *materiel* previously defined as weapon systems and major equipment items. Here the term *materials* indicates all of the components, bits, and pieces that are necessary to work on an item, plus the item itself. This material could be hydraulic actuator repacking kits, or hydraulic fluid, or sheet metal, or any material that is required in addition to the item being repaired. The final input to the maintenance process is "information." This information input is different from the procedures input. This input is information about the item to be repaired or maintained. This could be information about the condition of the item, e.g., is it inoperative or merely out of tune. Or it could be information about how the item failed, when the item failed, where it failed, how it failed, and so on.

The Maintenance Process combines and uses all of these inputs, and produces some "output." Depending upon the organizational level, or the environment, the specific output may be different for each Maintenance Process. In general, however, the output of this maintenance process is mission ready weapon systems and serviceable equipment. This output can also be expressed in some form of capability: combat capability, operational readiness, mission readiness, or other capability. These outputs should be focused on creating and sustaining some high level of warfighting capability. Hence, maintenance is a process whereby an item of equipment or weapon system, along with certain other inputs, are combined to produce a serviceable or combat capable item that supports operational requirements.

17.2   Historical Development of Maintenance

Air Force maintenance had its foundation in the beginning days of aviation. The first aircraft maintenance was that of the "owner/operator" kind. If the aircraft required maintenance it was the pilot's responsibility. However, it was not long before the complexity of the emerging aircraft technology required a more highly skilled maintenance concept. On 8 May 1913, the *U.S. Army Aviation Section Technical Order* 00-2A was published. This technical order established the first military version of the "crew chief."

After World War I, the trend toward maintenance specialization which had developed during the war began to change. The post war years were very lean years for both military and civilian aviation. As a result, the mechanics had to become generalists. The crew chief system became more formal due to the post war exodus of qualified mechanics. Teams of broadly trained mechanics were established and worked for an experienced master mechanic, the crew chief. The few specialists that existed were not members of the crew but were needed to maintain the more sophisticated equipment of the day: the armament, camera, and radio systems.

The years between the world wars saw many advances in the way aircraft maintenance was performed. The depot system was developed and, by 1938, four depots existed at San Antonio, Texas; Fairfield (now Fairborn), Ohio; Rockwell (now Sacramento), California; and Middletown, Pennsylvania. This depot system employed about 1,700 people and overhauled 166 aircraft and 500 engines per year. The total aircraft inventory in 1939 was less than 2,500 airplanes. During this time the inspection interval system evolved from the previous daily, weekly, or monthly bases to a flying hour basis.

The complexion of the aircraft maintained by the Army Air Corps changed. During World War II, at the onset of the war, the most advanced aircraft in the U.S. arsenal was the B-17. By the end of the war, the aircraft being maintained had substantially increased in sophistication. The B-29 was the newest member of the inventory and was a very complicated piece of flying hardware. In addition, jet technology was beginning to emerge. The result was a more complicated aircraft inventory which had to be maintained with a small, less technically qualified workforce. Some changes had to be made. It was the Strategic Air Command (SAC) which initiated those changes. In August 1949, SAC published SAC Regulation 66-12, which emphasized a high degree of technical specialization.

This specialist-oriented maintenance organization was managed by a strong, centralized control structure under the direction of a single, senior officer: the Chief of Maintenance. SAC, alone, used this concept for the next nine years.

The Korean War again tested the Air Force deployed maintenance capabilities to the limit. The specialist system of maintenance employed at this time did not produce the desired results in the early stages of the war. The heavy operational requirements, when coupled with crowded and inadequate in-country maintenance support, resulted in rapid deterioration of the aircraft. To counter this situation a system, known as Rear Echelon Maintenance Combined Operations (REMCOs), was established. The forward bases performed the launch, recovery, and servicing type of maintenance, while the rear bases performed the scheduled and heavy maintenance. This REMCO system allowed the forward based mechanics to focus on sortie generation while the rear-based maintenance specialists were able to concentrate on repair and overhaul of the aircraft.

By the end of the Korean War, other operational commands experienced the same lack of skills and increase in system sophistication that SAC had experienced at the end of World War II. Therefore, in July 1958, the Air Force adopted the Strategic Air Command's specialist-oriented, centralized control maintenance concept as its standard maintenance structure for all operational commands. Although the thrust of this historical perspective, to this point, has been upon aircraft maintenance, other types of maintenance were also evolving: missile maintenance, vehicle maintenance, ground communications maintenance, and simulator and training device maintenance, to name a few. The benefits that aircraft maintenance discovered in the AFM 66-1 centralized maintenance concept were also evident to the leadership of other maintenance organizations. Therefore, in 1960, the maintenance of our strategic missile force began to use the AFM 66-1 concept. Maintenance of simulators and training devices came under the AFM 66-1 concept in 1961. The AFM 66-1 centralized maintenance management concept was next accepted by both ground communications and, vehicles, in 1962 and 1963 respectively.

The next prolonged war environment the Air Force encountered was in Vietnam. The SAC bombing operations were centralized in places like Guam, Thailand, and the Philippines. The maintenance conditions in these safe sanctuaries were almost identical to the conditions that existed at SAC bases in the Continental United States (CONUS). Therefore, SAC maintenance operations continued to use the AFM 66-1, centralized maintenance concept. On the other hand, the tactical fighter aircraft used in Viet Nam needed to be responsive to the local battle conditions and were therefore located closer to the field of battle. When these tactical units were deployed to Viet Nam, they employed a variation of the AFM 66-1 maintenance concept. This variation was called the TAC Enhancement Program . Under this variation, the organizational maintenance squadron personnel were transferred to the control of the tactical fighter squadron commander. However, even this TAC Enhancement Program encountered problems when it was tasked to produce the numbers of sorties that the war in Viet Nam required. As a result of this shortcoming, the Air Staff began an effort to study ways of improving the maintenance support.

During the Yom Kippur War, in October 1973, the Israeli Air Force, flying the same types of aircraft used by TAC, was able to generate a remarkably high sortie rate. TAC representatives found that organizational and specialty differences were minimized in Israeli maintenance organizations. Maintenance specialists were assigned to the flight line, worked side-by-side with the crew chiefs and performed whatever tasks were necessary to launch and recover the aircraft. The emphasis in the Israeli maintenance organization was on generating a high sortie count, which was accomplished by minimizing the time it took to launch and recover each aircraft. The maintenance leadership was not hobbled with the problem of determining which individual, or specialist, could work a particular job; "everybody did everything." This difference in orientation, this lack of specialization, seemed to be the major difference in the effectiveness between the USAF and Israeli Air Force maintenance production.

By September 1974, the Tactical Air Command had begun developing and testing a variant of the Israeli maintenance concept with the expressed purpose of increasing the command's sortie generation capability. The result was known as Combat Oriented Maintenance Organization (COMO) (Multi-Command Regulation 66-5). In 1977 PACAF converted to COMO, and, by 1980, USAFE had also converted. The organizational change to COMO was made to expand total work force flexibility, simplify specialist dispatch, and decentralize production decisions to improve sortie capability.

Figure 17.1 depicts the COMO structure, while Fig. 17.2 shows the typical specialist oriented maintenance organization. Besides the obvious difference in the number of squadrons, the major difference between the specialist oriented organization and the COMO was in the Job Control function of Maintenance Control, or Maintenance Operation Center (MOC). Under the specialist oriented maintenance organization, job control was very powerful and directed and controlled all of the scheduled and unscheduled base maintenance actions as well as assigned all of the operational maintenance priorities. Under the combat oriented maintenance organization, job control became more of a "status keeper" with the real maintenance production and control responsibility pushed lower in the organizational structure (decentralized) to the Aircraft Maintenance Unit (AMU) Production Supervisors.

The job control status-keeping function still existed, but the "real time" job control function rested much lower in the organization.

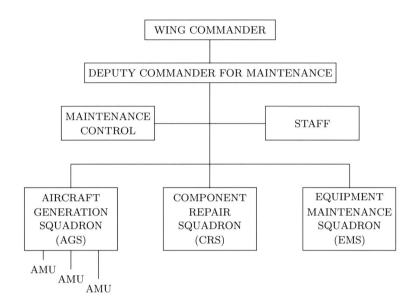

Fig. 17.1: Typical COMO maintenance organization.

Under COMO the responsibilities for the three maintenance production squadrons were delineated, not by specialist function, as in AFR 66-1, but by whether they were a direct sortie-producing element or an indirect sortie-producing element. The direct sortie-producing element consisted of the Aircraft Generation Squadron (AGS) while the indirect sortie-producing element was composed of both the Component Repair Squadron (CRS) and the Equipment Maintenance Squadron (EMS). Prior to COMO, 75 percent of the sortie-producing maintenance personnel were located in the shops of Field Maintenance Squadron (FMS), Avionics Maintenance Squadron (AMS), and Munitions Maintenance Squadron (MMS). COMO also distinguished between on-equipment maintenance and off-equipment maintenance. AGS was a direct sortie-producing element that performs on-equipment maintenance. CRS and EMS were indirect sortie-producing elements that performed off-equipment maintenance.

The Aircraft Generation Squadron (AGS) was created by taking the old flight branch of crew chiefs (from OMS) and adding the flight line specialists from FMS and AMS and the load crews and on-equipment weapons release

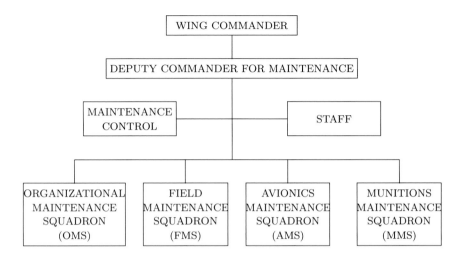

Fig. 17.2: Typical specialist oriented organization.

and gun services specialists from MMS. CRS was responsible for the off-equipment, intermediate level maintenance on avionics and aircraft systems. The off-equipment personnel from FMS and AMS were combined to form the CRS. EMS was responsible for the off-equipment, heavy repair intermediate level maintenance and munitions repair and storage. In addition, the major scheduled inspections were also the responsibility of EMS. EMS received the AGE, fabrication, and repair and reclamation technicians from FMS and the munitions maintenance and storage specialists from MMS. As with the CRS, personnel from EMS were usually not dispatched to the flight line to perform any on-equipment maintenance.

The COMO, or decentralized, maintenance concept demonstrated several advantages. The first advantage was in the area of sortie generation. When General Creech arrived at TAC in 1978, the sortie rate had been falling for ten years at a compound annual rate of 7.8 percent. From 1978 through 1983, it rose at a compound annual rate of 11.2 percent. A second advantage of COMO was associated with the first – the time lag between requesting a specialist, and their arrival at the aircraft, was reduced. By having functional specialists assigned to the AGS, specialist response time was shortened. A third advantage related to smoother work requirements. In specialist oriented maintenance individual specialist or shop workloads could be cyclical. Under COMO, the impact of high specialist work requirements could be dampened by other Cross Utilization Training (CUT) specialists. Likewise, when a particular specialist workload was low, these specialists could help other technicians. Therefore, under COMO, the mission, not the speciality, determined the workloading. A fourth advantage of COMO involved the level of decision making and control. Under specialist oriented maintenance, control was centralized in Job Control. In COMO the real control of the maintenance production rested with the AMU Production Supervisor. Therefore, the urgency of a maintenance action could be assessed on the spot and technician support requested from a workforce which is assigned and available within the AMU.

COMO also exhibited some limitations. When compared with the centralized maintenance organization, COMO required more people to accomplish the maintenance mission. The strength of COMO, unit identification and personnel allocation, created the need for more maintenance people. A second limitation was the increased specialization created by dividing equipment specialists into on- and off-equipment specialists. While any particular specialist became more proficient at a portion of their specialty, from a systems perspective, the specialist became more job knowledge limited. This second limitation also contributed to a third limitation. Although the maintenance personnel assigned to the Aircraft Generation Squadron exhibited high levels of "ownership" and mission orientation, what about the personnel assigned to the Component Repair Squadron or Equipment Maintenance Squadron? Since they were no longer directly associated with the flying mission of the wing, a tendency developed for these people to become more specialty-oriented and to lose their mission orientation. A final limitation was the increased demand that COMO placed upon the base support equipment resources. A base COMO structure that had three Aircraft Maintenance Units (AMU) required more support equipment than a single flight line oriented squadron. As a result, duplicate base level support equipment was required. Thus, the resultant increased sortie-generation claimed by COMO came at a cost of more base level resources.

To reduce the number of resources needed in terms of maintenance personnel, Rivet Workforce was implemented in 1987. The objective of this program was to combine related aircraft maintenance Air Force Specialty Codes (AFSCs) to increase the efficiency and effectiveness of aircraft maintenance. Combining related AFSCs increases the number of skills for the individual maintenance technician and also gives supervisors greater flexibility to manage today's decreasing workforce. The 180 maintenance AFSCs, including "shred-outs," that existed prior to Rivet Workforce have now been reduced to 100 and renumbered to a 45XXX, the "Airman Manned Aerospace Maintenance" career field. Therefore, besides maximizing the training and utilization of maintenance personnel, Rivet Workforce has created a more mobile and survivable workforce that better meets future employment concepts.

Rivet Workforce is the cornerstone of the decentralized maintenance concept currently employed for accomplishing aircraft maintenance by the objective wings in all major commands. The objective wing is a result of the restructuring that the Air Force has undergone since 1991. The idea behind the objective wing is for the Air Force to organize in peacetime like it fights in wartime. An objective wing may consist of one particular type of aircraft, which is termed a monolithic wing, or may be comprised of several types of aircraft, which is known as a composite wing. As shown in Fig. 17.3, under the objective wing, the deputy commanders for operations, maintenance, and resource management have been replaced with group commanders. The intermediate level maintenance and resource management functions are now the responsibility of the Logistics Group commander. The maintenance squadron in the Logistics Group consists of COMO's CRS and EMS squadrons. Furthermore, to give operational control of aircraft to the flying squadron commanders, the organizational level maintenance performed by AGS under COMO has been moved to the Operations Group. Thus, the new base maintenance organizational structure is quite similar to that employed in the TAC Enhancement Program in the late 1960s.

17.3 Maintenance Functions

Having discussed the inputs, outputs, and objective of the Maintenance Process, it is time to understand more about the Maintenance Process itself. Within the Maintenance Process there are three distinct elements or functions – Maintenance Engineering, Maintenance Production, and Maintenance Management.

In Fig. 17.4 the life cycle of a particular equipment item is represented across the horizontal axis, from left to right. The division of the equipment maintenance process vertically shows the relative involvement of two of the functions as a reparable equipment item progresses through its life-cycle. Early in the item life-cycle, Maintenance Engineering functions dominate the maintenance process, while later in the life-cycle, the Maintenance Production functions dominate. Note also that the Maintenance Engineering functions never completely go away. Maintenance Management is inherent in both of these other functions.

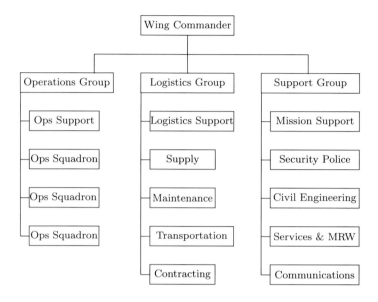

Fig. 17.3: Objective Wing structure.

MAINTENANCE ENGINEERING

Maintenance Engineering is the "technical" function of equipment maintenance. Its purpose is "to improve the reliability and maintainability of the equipment already being maintained and to ensure that optimum reliability and maintainability are designed into new equipment." Maintenance Engineering is responsible to see that the 1) Technical Requirements, 2) Maintenance Concepts, and 3) Equipment Designs are integrated to improve equipment logistical support as well as reduce support costs. Early in the acquisition process, Maintenance Engineering "provides concepts, plans, and data to define reliability, maintainability, and supportability requirements." By participating in the design process, and insuring that supportability requirements are met, the Maintenance Engineering function helps to provide the inherent levels of reliability and maintainability so that the production function of equipment maintenance, Maintenance Production, can have the item/equipment ready when it is needed.

The integration process begins early in the Conceptual Phase of the Acquisition Process and ends when the item is removed from service. Therefore, the responsibility of Maintenance Engineering occurs throughout the entire life cycle of an item. It begins early in system/item acquisition, as an

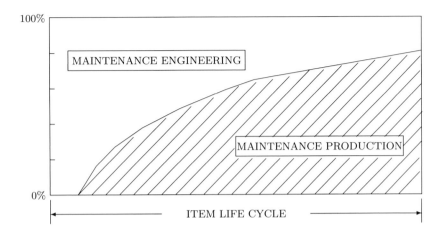

Fig. 17.4: Relative involvement in maintenance engineering and production.

identification of the technical requirements of the equipment design. These technical requirements are often identified by asking questions about data requirements and about what, when, why, how often, and by whom certain processes must be done. The results of these questions are examined in light of the "form," "fit," and "function" characteristics of the physical design, and a Maintenance Concept is specified. The specific Maintenance Concept evolves, over time, into a formal document called the Maintenance Plan.

The entire F-16 Maintenance Plan was only 50 pages long, and that included the index, USAF policies, and European Participating Governments interfaces. The Maintenance Concept is a product of the Maintenance Engineering responsibility, and specifically addresses requirements related to such issues as level of repair, modification processes, facilities, personnel, and training. This Maintenance Engineering information evolves, and is refined, over time. It is the responsibility of Maintenance Engineering to establish both the content and frequency of the maintenance that will be required and performed. To understand the complexity of the Maintenance Engineering task, one of the maintenance issues will be described – Level of Repair.

The Level of Repair decision is important because it impacts the allocation of workloads at both the bases and depots. There exist two basic Level of Repair models: the Three Level Model and the Two Level Model. Until just recently the three level model was the one predominantly used for most aircraft, engines, and component parts. Although the two level model is now the rule, particularly for all new weapon systems, it is useful to examine both models to better understand the rationale for implementing two level maintenance. As shown, in Fig. 17.5, the Three Level Model consists of the following levels of maintenance: organizational, intermediate, and depot.

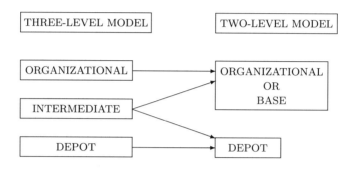

Fig. 17.5: Level of repair.

The Level of Repair decision specifies the organizational level where a particular item will be repaired. This decision will either (1) generate a new repair capability, or (2) take advantage of an existing repair capability. It is an important point to realize that the repair of a particular item can be either (1) entirely supported at a single level (say at the Depot Level) or (2), divided between several levels of repair. For instance, in making the decision as to where to repair a landing gear strut, the Organizational Level could be required to inspect the strut and perform routine fluid-level servicing. Likewise, the Intermediate Level could be required to perform minor repair, such as removing and replacing the "O" rings and seals. Finally, the Depot Level could be assigned all other major repair of this landing gear strut. Therefore, the Level of Repair decision is one which "allocates" all of the maintenance actions to one, or more, Levels of Repair.

The Organizational Level is frequently synonymous with the "operator" level, or "flight line," if specifying an aircraft system or item. Traditionally, the maintenance tasks performed at this level are usually only the "minor" maintenance tasks, or inspection, cleaning, lubrication, servicing, and minor adjustments. The usual argument for NOT placing more complicated maintenance at this level is the expense of placing skilled technicians or craftsmen at many dispersed sites or bases. The relatively low maintenance skill levels required, at this level of repair, allow the maintenance technician to use checklists, or abbreviated technical procedures. While more detailed technical procedures may be used, the traditional maintenance tasks performed at this level are usually preventative in nature.

The second level in the Three Level Model is the Intermediate, or Field Level. This maintenance level is traditionally associated with shop or "specialist" type maintenance. The maintenance frequently performed here is more involved than the maintenance performed at the Organizational level. This level usually employs more highly-trained personnel who perform component repairs. Some of the maintenance tasks might include adjustment, repair, extensive inspection, testing, and rebuilding of certain components. This more skilled work frequently requires extensive technical procedures and documentation. The testing and inspections performed at this level frequently require special tools, equipment, and test stations. The more complex work often requires more exotic special tools and support equipment. The F-16 Automatic Test Station is an example of a very expensive and complex piece of test equipment that is required for Intermediate Level maintenance. This Intermediate Level of maintenance may be located at each base or it may be consolidated at one location to support the Intermediate Level maintenance needs for several bases.

The final level in the Three Level Model is the Depot Level. The type of maintenance performed at this level is frequently at the "industrial" level. The tasks performed at this level include: major modifications, alterations, inspections with disassembly, and retrofitting. Some of the maintenance performed at this level may be one-of-a-kind, or state-of-the-art, type maintenance. This level of maintenance frequently requires extensive maintenance training, and even engineering skills. Therefore, the requirement for technical procedures and documentation is extensive, and frequently informal. Some of the maintenance performed at this level may not have been anticipated and, therefore, technical documentation may not exist. The contractor who developed the item may become involved with maintenance at this level. The support equipment, and tooling requirements at this level, can be exotic, and frequently may be one-of-a-kind machinery or test stations. The Depot Level therefore possesses an enormous industrial capability in order to perform maintenance.

In the Three Level Model, an item can be repaired at the Organizational Level, the Intermediate Level, or the Depot Level, or some level of repair can be assigned to each of the levels. An example of a decision to perform all maintenance at a single level concerns the multi-layer printed circuit boards used on the USAF Satellite Communication System. Because the test equipment required to isolate the problems in these printed circuit boards was so expensive, and the technical skill requirements were so extensive, the initial level of repair decision was made to repair ALL of these circuit boards at the Depot Level. Or the level of repair decision could require maintenance to be performed at all three levels. Given this decision, the Organizational Level mechanic could troubleshoot an item (fault-isolate the trouble to a particular item), then the maintenance specialist from the Intermediate Level could remove the item from the aircraft and troubleshoot the specific fault, down to an item or component, the latter which would then be shipped to the Depot Level for actual repair actions.

The Two Level Model is a variation of the Three Level Model. In the Two Level Model, the trouble shooting and repair responsibilities of the Intermediate Level have been separated into "on-equipment" and "off-equipment" tasks. The on-equipment tasks become the responsibility of the Base Level and the off-equipment tasks become the responsibility of the Depot Level. The assignment of actual repair capability in the Two Level Model reflects the same kinds of assumptions that were used with the Three Level Model. That is to say, there are trade-offs that consider the complexity of the maintenance actions, the skill levels of the technicians, and the complexity of the technical procedures, documentation, and support equipment requirements.

A decision to repair a particular item or category of end items at base level, rather than at the depot, means that base supply will require more personnel, more warehouse space, and more inventory in terms of end items and bits and pieces (i.e., bench stock) to repair these end items. For base maintenance, more facilities and test equipment will be needed, as well as more personnel with higher skill levels. Therefore, in terms of facilities and test equipment required, three level repair is generally more expensive than two level repair, especially when one factors in the multiple bases involved.

In 1992, with defense budgets declining and the reliability of newer aircraft, engines, and component parts increasing, the decision was made to implement two level aircraft maintenance across the Air Force. The only aircraft exempted from this concept were those in black programs, those that were few in numbers, and those close to being retired from the active inventory, like F-4s and F-111s. For commodities, the major factors examined in the two level decision were the cost of intermediate level support/test equipment and the number of intermediate level maintenance technicians required. Using these criteria, the majority of engines and avionics components became the first candidates for two level maintenance. The primary avionics components excluded were those with extremely low reliability. For engines, some residual Jet Engine Intermediate Maintenance (JEIM) capability to change fans, augmentors, and gearboxes and to perform emergency inspections was retained at base level. Also kept at this level were such intermediate maintenance functions as aircraft washing, egress maintenance, fuels maintenance, nondestructive inspection (NDI), and sheet metal repair and fabrication. In summary, two level maintenance not only reduces maintenance costs but also simplifies the Air Force's ability to deploy by eliminating much of the intermediate level test equipment formerly required.

Other maintenance issues that have to be addressed under the Maintenance Engineering function include inspection frequency, procedures and tools, economic repair limit criteria, and support equipment requirements. The maintenance work content and frequency decisions that are made in this portion of the Maintenance Process must be realistic, for together with the actual equipment design, they establish the "inherent" reliability and maintainability of each particular weapon system or equipment item. It is this "inherent" reliability and maintainability that is "passed" to the second function in the Maintenance Process, "Maintenance Production."

MAINTENANCE PRODUCTION

Maintenance Production is that function of the Maintenance Process where the actual maintenance is performed. While Maintenance Engineering is technical in nature and can be performed using analytical techniques, Maintenance Production is work oriented and occurs wherever maintenance actions are performed. Therefore, Maintenance Production occurs at each and every REPAIR level. It is oriented toward providing serviceable equipment that is available for use. The technical requirements and operational experience dictate the maintenance tasks for each level of maintenance. Maintenance Production is broken down into two categories of maintenance: On-Equipment and Off-Equipment. This division of maintenance provides a convenient way to classify where the maintenance actions are taking place. "On-Equipment" maintenance is performed on end-items of equipment, including engines, while "Off-Equipment" maintenance is performed on components that are removed from end-items of equipment, which are generally processed through maintenance or repair shops.

The combined activities of On- and Off-Equipment maintenance provide the Process of Maintenance with a feedback loop. Maintenance Production activities typically address questions regarding frequency of repair, repeat failures, length of repair time, cost to repair, length of time an item is out of service, and reliability projections. The answers to these questions are fed back to the Maintenance Engineering elements. The "demonstrated," or "achieved" reliability and maintainability (R&M) of the weapon systems, or equipment items, in the field is fed back and compared with the "inherent" R&M which was projected by Maintenance Engineering. If the two sets of figures agree, then the Maintenance Process is considered to be in balance. This means that the Maintenance Process is achieving the programmed, or inherent, R&M characteristics of the subject weapon system or equipment. If, on the other hand, the field data (the "achieved" R&M) do not match the "inherent" R&M, then a system "imbalance" exists. The reasons behind this "imbalance" must be found, and analyzed, by the Maintenance Engineering portion of the process.

The Maintenance Process might be out of balance due to inadequate or incorrect repair procedures, inadequate tools and test equipment, inadequately trained and/or supervised maintenance personnel or inadequate supply support. In addition, equipment design that is not as maintainable in the field or as reliable as originally predicted may cause the process to become unbalanced.

When a Maintenance Process is out of balance, this usually results in some technical improvement and modification to hardware, software, and/or support data for the subject weapon system or item of equipment. That modification will typically involve either a physical change in the equipment design, or a change in the maintenance procedures. These changes subsequently improve the reliability and maintainability of weapon systems and equipment. Consequently, maintenance engineering efforts play a significant role in establishing and controlling the requirements for maintenance resources.

MAINTENANCE MANAGEMENT

The third function of the Equipment Maintenance Process is Maintenance Management. Maintenance Management is the application of the five classical management functions of planning, organizing, coordinating, directing, and controlling to the other two functions of equipment maintenance – Maintenance Engineering and Maintenance Production. This function recognizes the importance of quality management at all levels of maintenance planning, policy, coordination, and production. It therefore includes the management actions that occur at all the various organizational levels within the U.S. Air Force, including HQ USAF and the operating commands. The importance of this function is that it highlights the critical nature of quality management actions surrounding the Equipment Maintenance Process. Whether developing policies, procedures, or programs; or monitoring the accuracy and timeliness of maintenance actions; quality maintenance management is both necessary and required. Since this management is inherent in the other two elements or functions of equipment maintenance, it will not be discussed further but will be assumed to be an integral part of the entire Maintenance Process.

17.4   Maintenance Tasks

CLASSICAL TASKS

Maintenance tasks are the physical actions taken by maintenance personnel to keep equipment serviceable, safely operable, properly configured, and available for use. The three classical maintenance tasks are prevention, correction, and modification. An effective Equipment Maintenance Process must "prevent" equipment failure or malfunction and does this by periodically servicing or inspecting equipment. The Maintenance Process must also "correct" malfunctions or physical damage by repairing or rebuilding the equipment. Finally, the effective Maintenance Process must "modify" equipment when defects occur, safety is involved, or performance standards must be changed.

The first task, "Prevention," is best described as the scheduled work that is performed to avoid failures. Preventive maintenance actions include cleaning, lubrication, inspections, visual checks ("walk-arounds"), recalibration, scheduled removal and replacement, minor adjustments, and corrosion control. The objective of a preventive maintenance program is to realize the inherent reliability of the equipment, or said more simply, to "ensure safe and reliable operation of the equipment." This program includes the following four types of tasks:

1. Scheduled inspection of an item at regular intervals to find any potential failures;

2. Scheduled rework of an item at, or before, some specified age limit;

3. Scheduled discard of an item (or one of its parts) at, or before, some specified life limit; and

4. Scheduled inspection of a hidden-function item to find any functional failures.

With the implementation of the Reliability Centered Maintenance (RCM) concept, the Preventive Maintenance task has become increasingly more important in recent years. This concept attempts to eliminate problems before they become system or item failures.

Corrective Maintenance, on the other hand, is work that is performed *after* failures have occurred. This is maintenance that is performed to restore an item to satisfactory condition after a failure or malfunction has caused degradation of the item below the specified performance level. Some of the specific maintenance tasks that are usually associated with Corrective maintenance include troubleshooting; disassembly; adjustment; repair;

remove and replace; remove, repair and replace; and re-calibration. Much of the Corrective maintenance is performed Off-Equipment, or in the maintenance shops, where test equipment or special tools are available to the mechanic. From a supply standpoint, it is the Corrective maintenance tasks which repair the failed items that are turned in to the supply system, requiring maintenance. After this Corrective maintenance, the repaired item is returned to the supply system for future use.

Modification is necessary when equipment needs to be changed to better serve its intended purpose. Some of the reasons to modify an item include safety considerations, new use for equipment, reliability is too low, high maintenance costs (time, personnel, etc.), new technology, and change in mission. Item managers, maintenance managers, and commanders must be constantly aware of the mission needs and how their assigned equipment, in its current configuration, fits those mission needs. A Modification program highlights the need for some form of "configuration management." It is very important for the item manager to know the "configuration" of all of his items. This includes both installed items and items in stock. Imagine the problems that would be created if a weapon system required a modified item, yet not all of the items in the supply system had been modified. Therefore, the management of the Modification program greatly impacts both the maintenance effort and the combat mission of that equipment. Some of the specific maintenance tasks frequently associated with Modification maintenance include alteration, modification, retrofit, rebuild, remanufacture, and reclamation.

CONSEQUENCES OF EQUIPMENT FAILURES

There are several reasons why Maintenance is so important to the military. These reasons are based upon the consequences of not performing the proper maintenance at the correct time. There appear to be four consequences of not performing maintenance at the proper time – safety consequences, operational consequences, the direct cost of repair, and compound failure consequences.

The first consideration in evaluating any possible equipment failure is "Safety." This means the possible loss of equipment and/or its occupants. Suppose the failure in question is the separation of a number of turbine blades on an aircraft engine, causing the engine to vibrate heavily and lose much of its thrust. This failure could certainly affect the safety of a single-engine aircraft and its occupants, since the loss of thrust will force an immediate landing regardless of the terrain below. Furthermore, if the engine is one whose case cannot contain ejected blades, the blades may be thrown through the engine case and cause unpredictable, and perhaps serious, damage to the plane itself. Regular inspection for this condition (damaged turbine blades, or stress fatigue symptoms) as part of scheduled maintenance makes it possible to remove engines at (or before) the potential-failure stage, thereby forestalling all critical functional failures. Therefore, a well-designed inspection and replacement program can preclude serious loss of equipment and occupants.

The second consideration in evaluating any possible equipment failure is the "Operational consequences." Whenever the need to correct a failure disrupts planned operations, the failure has operational consequences. Thus, operational consequences include the need to abort an operation after a failure occurs, the delay or cancellation of other operations to make unanticipated repairs, or the need for operating restrictions until repairs can be made. If a potential failure such as loose turbine blades were discovered while the plane was in service, the time required to remove this engine and install a new one would involve operational consequences. However, inspections for this potential failure can be performed while the plane is out of service for scheduled maintenance. In this case there is ample time to remove and replace any failed engines without disrupting planned operations. A well conceived inspection program that can identify potential equipment failures and is integrated with routine servicing and quality repair work [very few repeat or recurring failures] can preclude failures that impact mission accomplishment.

These first two consequences have definite operational consequences, but there are many kinds of functional failures that have no direct adverse effect upon the operational capability. One common example is the failure of a navigation unit in a plane equipped with a highly redundant navigation system. Since other units ensure availability of the required function, the only consequence in this case is that the failed unit must be replaced at some convenient time. Thus the costs generated by such a failure are limited to the cost of corrective maintenance. This navigation unit may have to be repaired as an "unscheduled" maintenance action, instead of the repair being "scheduled." The more "unscheduled" or "emergency" maintenance that is performed, the less resources you have to dedicate to "scheduled" or Preventive maintenance actions. Therefore, these non-operational failures can extol some "opportunity cost" in addition to the direct cost of repair.

The fourth and final consequence of failure is another non-operational consequence, the problem with "hidden," or compounding failures. These are failures that have no immediate consequences, i.e., no direct adverse effects. However, these "hidden" failures, if not detected and corrected, can ultimately create a major problem or failure. For example, certain elevator-control systems are designed with concentric inner and outer shafts so that the failure of one shaft will not result in any loss of elevator control. If the second shaft were to fail after an undetected failure of the first one, the result would be a critical failure and could impact safety. In other words, the consequence of any hidden-function failure is increased exposure to the consequence of a multiple (or compound) failure. One way to address the problem of "hidden," or compound, failures is the use of "redundant" systems and scheduled inspections.

17.5   Depot Support

The Depot Level maintenance support provides all three of the Equipment Maintenance functions of Maintenance Engineering, Maintenance Production, and Maintenance Management. Within the U.S. Air Force, this Depot Level support is provided by the Air Force Materiel Command (AFMC). This section will only discuss the maintenance related functions of AFMC and, more specifically, only those functions that relate directly to Depot Level maintenance.

MAINTENANCE ENGINEERING FUNCTION

Depot capability assessment is performed for every item, or weapon system, that potentially requires depot level maintenance support. This assessment includes an analysis of the requirements for organic (in-house), interservice, and contractor depot support. If another service already possesses organic depot support, the Air Force and that service would evaluate the impact that this new maintenance workload would impose upon it. Usually existing depot-level maintenance capabilities are not duplicated. For this reason, all depot-level maintenance for the Air Force UH-1 helicopters is performed by the Army. It is the Air Force policy to achieve maximum use of interservice support when this support results in greater economy or effectiveness.

If an organic Air Force maintenance capability is required, AFMC will assign that requirement, in accordance with the Technology Repair Center (TRC) concept. Implemented in 1974, the TRC concept insures the concentration of technical expertise and advanced repair capability at the correct locations within the AFMC command. The TRC concept assigns homogeneous work loads to a single Air Logistics Center (ALC), based upon technological requirements. For instance, all aircraft landing gear are repaired at Ogden ALC. This centralized repair concept establishes a depot repair expertise and then repairs all like items at that particular location. The TRC concept replaced the area, or geographic, depot level maintenance responsibilities previously administered by the Air Material Areas (AMAs).

For economic, timing, or surge capability reasons, the Air Force may choose to use contractor depot level support. Contractor support is frequently an effective way of augmenting in-house, or organic, depot level maintenance support. Flexibility, survivability, responsiveness, capital investment, and workload leveling are just a few of the advantages that can accrue from using contractor depot level maintenance. Cost is always an important consideration, when evaluating the contracting capability alternative. The existence of this form of support also helps keep the Air Force organic support costs down. For, when depot level work is being planned, the ALC with primary responsibility must submit a "bid" for that support. This bid will be compared with any contractor bids before a decision is made. If an ALC is not competitive with similar contractor support, the ALC may not be assigned the maintenance workload. Therefore, cost-competitive ALC maintenance organizations are rewarded with new maintenance workloads. Long range workload balancing between the ALCs is resolved, using the AFMC Depot Maintenance Posture Plan .

MAINTENANCE PRODUCTION FUNCTION

AFMC is responsible for performing the depot level maintenance production functions. Although Headquarters AFMC is involved, the vast majority of this maintenance production activity is administered by the Air Logistics Centers (ALCs). After a short description of ALC organizational structure, we will look at three depot level maintenance production activities: programmed depot maintenance, combat logistics support, and reclamation.

Figure 17.6 is a map that identifies the major ALCs. AFMC headquarters is located at Wright-Patterson AFB, OH. The five ALCs are Oklahoma City ALC, Ogden ALC, Sacramento ALC, San Antonio ALC, and Warner Robins ALC. Each ALC, with the exception of Sacramento, is the largest single employer in its respective state. In addition to the five ALCs, there are several specialized organizations which support the AFMC depot maintenance mission. The Aerospace Guidance and Metrology Center (AGMC) in Newark OH performs depot level maintenance for aircraft and missile guidance systems as well as calibrating the Air Force Precision Measurement Equipment. The Aerospace Maintenance and Regeneration Center (AMARC) in Tucson AZ performs depot-level storage, reclamation and disposal activities for the DoD. AMARC was formerly named the Military Aircraft Storage and Disposition Center (MASDC) and was affectionately called the "bone yard." The name change from MASDC was recognition that AMARC performs a significant regeneration mission. Since 1968, more than 3,000 aircraft and drones have been returned to the active inventory or used for the Security Assistance Program. While not actually performing depot level maintenance, the Cataloging and Standardization Center

(CASC) in Battle Creek MI, assists the maintenance effort by establishing and maintaining the Air Force item identification and supply management data.

AGMC - Aerospace Guidance & Metrology Center
AMARC - Aerospace Maintenance & Regeneration Center
CASC - Cataloging and Standardization Center

Fig. 17.6: Map of Air Logistics Centers and associated units.

Programmed Depot Maintenance (PDM) is a scheduled maintenance program that occurs throughout the life-cycle of each aircraft, engine, or major end item. Periodically, most Air Force aircraft are returned to their prime ALC where a specified work package is performed. A typical package may accomplish some major Time Compliance Technical Orders (TCTOs), any outstanding major modifications, and a corrosion control treatment, including new paint. A TCTO usually involves a one time change, modification or inspection of installed equipment or systems. The specific contents of the PDM work package are negotiated between the operating commands and the maintaining, or supporting, ALC. The contents of the PDM work package may be affected by the results of Analytical Condition Inspections (ACIs), the age and condition of the specific aircraft, and the economic and operational environments at the time. The ACI is a major inspection of a particular type of aircraft performed at its supporting ALC. This inspection is performed on only a few aircraft and provides the depot level maintenance experts with extensive knowledge about structural, internal component, corrosion, and potential fatigue problems. In contrast to the inspect and repair as necessary (IRAN) depot maintenance of a decade or so ago, the PDM philosophy is designed to minimize the aircraft down-time and only repair those items and components that are necessary to return the aircraft back to a combat-ready condition. The average PDM costs about 15 to 20 percent of the aircraft "stock" price. Some of our newer fighter aircraft did not program for a PDM program. Instead, these aircraft use a "Speed Line" depot scheduled maintenance program that is very similar to the PDM concept.

Combat logistics support is an important, but little known, depot maintenance activity. Through five active duty and six reserve Combat Logistics Support Squadrons (CLSS), AFMC provides world-wide depot level maintenance support. While chartered to provide highly trained, world-wide deployable aircraft battle damage repair (ABDR) and combat distribution augmentation teams to operating commands in wartime, these CLSS also provide valuable field level support during peacetime. Often called depot field teams, these depot maintenance experts are frequently called upon to perform maintenance beyond the technical or manpower capability of operating bases. One example was the A-10 aircraft gas gun modification, where CLSS teams helped bases in two theaters perform this Time Compliance Technical Order (TCTO) in place. When not deployed to major combat exercises, or in training, the CLSS members augment the ALCs depot maintenance production staff.

Reclamation is unique to depot level maintenance. AFM 67-1 defines "reclamation" as a process of disassembly of excess aircraft, engines, and other end-items which recovers serviceable, or economically reparable, spare parts when requirements still exist. Except in some rare instances, such as crash damaged aircraft, reclamation efforts are directed and performed by depot level personnel. When an AFMC Item Manager (IM) or System Manager identifies a need, depot level maintenance personnel perform a reclamation action. Sometimes this request is in the form of a weapon system "save list," sometimes it is a one time occurrence to fulfill a priority, or Mission Capability (MICAP), request. In every instance, the reclamation effort performed by depot level maintenance is significant. In 1984, more than 50,000 aircraft parts were reclaimed from aircraft stored at Aerospace Maintenance and Regeneration Center (AMARC) in Tucson AZ. While some of these reclamation actions were routine "saves," about 20,000 parts were priority requests to support active inventory aircraft. One of the "parts" was a B-52 wing. The value of the parts reclaimed totaled $163 million. These were parts that did not have to be procured. Another kind of reclamation does not involve parts but entire aircraft. In 1984, about $228 million in aircraft were returned to service with the military, federal agencies, and other tax supported organizations. Besides spare parts and aircraft, scrap metal and other recoverables are also reclaimed at AMARC. If 1984 was representative of the peacetime reclamation effort, the value of reclamation during wartime should become obvious.

17.6   Maintenance Decision Impacts

Table 17.1 highlights some of the tradeoffs that are inherent in the Equipment Maintenance Process. The Maintenance Engineering decisions have been listed down the left-hand column. Across the top of the chart are listed some of the areas that the maintenance decisions impact or result in consequences. Due to size constraints, some other areas of impact are listed at the bottom of the chart. This list could contain a great number of impact areas, each of which is, or would be, affected by the maintenance decisions on the left-hand side of the chart. The purpose of the chart is to graphically depict the far-reaching impacts of the decisions made in the Equipment Maintenance Process. Table 17.1 emphasizes the fact that the Maintenance Engineering decisions are not "maintenance only" kinds of decisions. Instead, the impact-complexity that each maintenance decision can have should now be more evident. Consider the potential impacts of maintenance procedures and tools.

Table 17.1: Maintenance Management Decisions and Trade-Offs: Issues and Impacts/Consequences

ISSUES	IMPACTS/CONSEQUENCES
Level of repair.	Sorties.
Type of support.	Mobility.
Inspection frequency.	Survivability.
Procedures and tools.	Labor costs.
Economic repair limit.	Material costs.
Modification process.	Transportation costs.
Facilities.	Demonstrated reliability.
Personnel numbers.	Maintainability.
Personnel skills.	Base self-sufficiency.
Support equipment.	Surge capability.
Training criteria.	Availability.
	MIS support.
	Sustainability.
	Computer support.
	Supply support.

How do the decisions about the maintenance procedures and tools impact sortie generation? Here are some of the potential impacts:

- Lack of proper Procedures can impede the generation of Sorties because of inconsistent quality of maintenance, longer maintenance repair times, poor skill level training, missed inspection items, and improper fault isolation.

- Lack of proper Tools can impede the generation of Sorties because of damage to equipment parts, improper tolerances maintained, increased tool waiting time, and unmotivated maintenance personnel.

- Lack of proper Procedures can impact the unit's Mobility mission by making it impossible to support multiple locations.

This partial list should provide some insight into how complex and highly interdependent maintenance really can be. While the potential impacts used in the above example dealt with the "lack" of proper Procedures and Tools, the complimentary situation also creates impacts and consequences. Possessing the proper Procedures and Tools can "improve" the unit's sortie generation capability.

References

[1] Department of the Air Force, The U.S. Air Force Equipment Maintenance Program, AFR 66-14, Washington DC: HQ USAF, 15 December 1986.

[2] Department of the Air Force, Combat Support Doctrine, AFM 1-10, Washington DC: HQ USAF, 1 April 1987.

[3] Wyatt, Milton R. and Carroll M. Staten, "Maintenance in the U.S. Air Force," In Logistics Management LOG 224 , Volume I, pp. 17.23-17.46, Edited by Dennis L. Hull and Albert H. Rogers, Wright-Patterson AFB, OH, Air Force Institute of Technology, 1985.

[4] Peppers, Jerome G., Jr., "History of Maintenance in the Air Force," Wright-Patterson AFB OH, Air Force Institute of Technology, December 1967.

[5] Townsend, James N., "A History of Aircraft Maintenance in the Army Air Force and the United States Air Force," Unpublished research study, Air Command and Staff College, Maxwell AFB AL, May 1978.

[6] Anderton, David A., "POMO and POST: Keystones of TAC Readiness," Air Force Magazine (January 1979), pp. 46-50.

[7] Peters, Tom and Nancy Austin, *A Passion For Excellence: The Leadership Difference*, New York, Random House, 1985.

[8] Foster, Dwight J. and John C. Olson, "A Comparative Evaluation of the Effects of the Implementation of the Production Oriented Maintenance Organization (POMO) on Aircraft Maintenance," Unpublished master's thesis, LSSR 27-78B, AFIT/LSGR, Wright-Patterson AFB OH, September 1978.

[9] The Inspector General of the Air Force, "Report on Functional Management Inspection of Rivet Workforce Transition and Utilization Program," 3 December 1991.

[10] McPeak, Merrill A., Briefing to the Maintenance Officer Association Conference, 1 November 1991.

[11] Nowland, F. Stanley and Howard F. Heap, "Reliability-Centered Maintenance," Office of Assistant Secretary of Defense (Manpower, Reserve Affairs and Logistics), Washington DC, 29 December 1978.

[12] Leachman, William D. and Thomas E. Vititio, "Maintenance Management," in Managing the Air Force , Fourth Edition. Maxwell AFB AL, Air War College, 1983, pp.14-1 - 19.

[13] Murray, Steve, "Depot Maintenance Extending Aircraft Service Life," Air Force Logistics Command News Service, Release Number 86-7-183, Office of Public Affairs, Wright-Patterson AFB OH, July 25, 1986.

[14] Diener, David A. and Barry L. Hood, "Production Oriented Maintenance Organization: A Critical Analysis of Sortie-Generation Capability and Maintenance Quality," Unpublished master's thesis, LSSR 52-80, AFIT/LSGR, Air Force Institute of Technology, Wright-Patterson AFB OH, June 1980.

Chapter 18

ENVIRONMENTAL ISSUES IN WEAPON SYSTEMS ACQUISITION

Lt. Col. Mark N. Goltz,
Lt. Col. Michael E. Heberling,
and Capt. James F. Donaghue
Air Force Institute of Technology

18.1 Introduction

In the past decade, the impact of environmental considerations on decision making within the Department of Defense has grown from a relatively insignificant side-issue to a major factor in the decision making process, commensurate with safety, cost, and operational considerations. Secretary of Defense Cheney has asked that the Department of Defense posture itself as the Federal leader in environmental compliance and protection, with "the first priority of our environmental policy...to integrate and budget environmental considerations into our activities and operations."[1] General Merrill McPeak, the Air Force Chief of Staff, echoed Secretary Cheney's sentiments with specific environmental goals for the Air Force. These goals range from restoration of hazardous waste sites created in the past, to complying with current laws and preventing pollution in the future.[2]

Decisions made during concept exploration and definition, demonstration and validation, engineering and manufacturing development, and the production phases of the acquisition process directly influence the cost of managing material and waste streams for production contractors and subcontractors, depots, and installations that maintain and operate the systems. They also affect the cost to the Air Force of disposing of the system at the end of its useful life. Decisions made during the acquisition process leave a legacy that lasts long after the acquisition is completed.

Over the past five years, the Air Force has experienced a tenfold increase in the unit cost of disposing of hazardous waste. Federal, state and local regulators now have the ability to inspect and fine our Government Owned Contractor Operated (GOCO) facilities and Air Force installations for environmental violations. Half of all Air Force violations are for mismanagement of hazardous materials and waste. Approximately 90% of the entire Air Force hazardous material and waste stream is the direct result of depot and field maintenance procedures prescribed in technical orders (TOs). Materials and processes prescribed in TOs are dictated by decisions made during acquisition. There is a direct cause and effect relationship between the technology, material and manufacturing process decisions made during acquisition, and the materials and processes required to maintain the systems. Making optimum technology decisions requires looking beyond the short term to their life cycle consequences; it requires a new way of thinking about the consequences of decisions made during acquisition. The Air Force may easily wind up paying many times over for a decision made during development in order to save a little in unit production cost.

Even as technology and process decisions are being made today for tomorrow's systems, the logistics community is reevaluating processes used to maintain existing systems because the cost of managing their waste streams is too high. Not only does it cost more to change a technology or process decision after the system is fielded, but the benefit of the change accrues over only a portion of the lifetime of the system. It is much more cost effective to select the correct technology or process before the system goes into production. As Operations and Maintenance (O&M) dollars decrease and the cost of managing pollution increases, selecting environmentally preferable technologies becomes ever more important if we are to secure the maximum warfighting capability at least cost.

To start the Air Force down this path of integrating environmental performance into the acquisition decision making process, the Secretary of the Air Force, Donald B. Rice, and the Chief of Staff, General Merrill A. McPeak, proposed a Pollution Prevention Action Plan on November 13, 1991.[3] This was followed by additional memoranda on the Air Force Pollution Prevention Action Plan and the Air Force Ban on Purchases of Ozone Depleting Chemicals, both dated 7 January 1993.[4,5]

The emphasis of this chapter will be on pollution prevention, using the Air Force program as an example of how pollution prevention concepts and techniques may be integrated into the weapons system design and acquisition process to prevent pollution before it occurs.

We will begin the chapter by defining pollution prevention and describing Department of Defense and Air Force policy as it relates to prevention of pollution. We will then discuss programs which are in place to incorporate environmental concerns into weapon systems acquisition. Then we will examine aerospace industry and Air Force initiatives and successes in reducing pollution in the acquisition process. In the final section of the chapter, the opportunities and challenges of the future, with regard to incorporation of pollution prevention into weapon systems acquisition, will be discussed.

18.2   Pollution Prevention

On October 27, 1990, Congress passed the Pollution Prevention Act of 1990. As part of the findings of that Act, Congress stated that there were significant opportunities to reduce or prevent pollution through source reduction techniques such as changes in production, operation, and raw materials. Congress also stated that source reduction was more desirable than pollution management or control. The Act focussed very strongly on source reduction, with recycling being both a less emphasized and a less preferred alternative to source reduction. Pollution that cannot be reduced at the source or recycled should be treated if feasible, and as a last resort, disposal into the environment should be conducted in an environmentally safe manner. Fig. 18.1 summarizes this hierarchy of options.

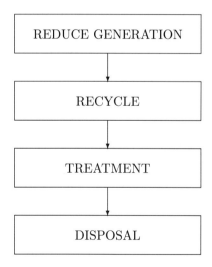

Fig. 18.1: Hierarchy of waste management options.

The Environmental Protection Agency (EPA) summarized the Act to the regions on November 27, 1990. The EPA stated that although there was no definition of pollution prevention contained in the Act, "it is clear that Pollution Prevention is intended to include source reduction and to exclude most forms of recycling. Pollution Prevention is not limited to hazardous waste or chemicals subject to (Toxic Release Inventory) reporting, but encompasses any hazardous substance, pollutant or contaminant."[6]

Source reduction is defined in the EPA memorandum as "any practice which reduces the amount of any hazardous substance, pollutant or contaminant entering the waste stream or otherwise released into the environment (including fugitive emissions) prior to recycling, treatment or disposal."[6] The practice must reduce the hazards to public health or the environment associated with the release. According to the Act, source reduction "includes equipment or technology modifications, process or procedure modifications, reformulation or redesign of products, substitution of raw materials, and improvements in housekeeping, maintenance, training, or inventory control."[6]

On November 13 1991, General Merrill McPeak, Chief of Staff of the Air Force, and Donald Rice, Secretary of the Air Force, sent a letter to all major commands with a proposed pollution prevention action plan. The overall goal of Air Force pollution prevention policy is to prevent future pollution by reducing hazardous material use and releases of pollutants into the environment to as near zero as feasible.

In order to meet that goal, several objectives were announced:

1. Reduce the use of hazardous materials in all phases of new weapons systems from concept through production and deployment and ultimate disposal–find alternative materials and processes, and measure their life cycle costs.

2. Reduce the use of hazardous materials in existing (deployed) weapons systems by finding less hazardous materials and processes and integrating them into TOs, MILSPECS and MILSTDS.

3. Reduce hazardous materials use and waste generation at installations (civil engineering, vehicle and aircraft maintenance, family housing, etc.) and Government Owned Contractor Operated (GOCOs) facilities.

4. Acquire world class pollution prevention technologies, and distribute them throughout the Air Force.

5. Apply new technology to pollution prevention; searching outside sources first, and conducting Air Force research where no alternatives exist.

6. Establish an Air Force investment strategy to fund the Pollution Prevention Program.[3]

To meet the objectives, Air Staff has published Air Force Policy Directive 19-4, which establishes Air Force pollution prevention policy. The policy follows the hierarchy of options in the Pollution Prevention Act of 1990, i.e., source reduction, recycling, treatment, and disposal as a last option. Pollution prevention efforts are intended to be far reaching, including not only hazardous material use and generation, but also medical waste, air emissions, water discharges (both point and non-point source), and municipal solid waste. Where possible, the Air Force will recycle municipal solid waste, use clean, efficient energy sources and transportation, and establish affirmative procurement procedures for products made with recovered materials. The policy indicates that pollution prevention is an Air Force wide program, including all organizations and activities, and that training will be provided to better accomplish the program goals.[7]

We will now look more closely at Objective 1, which deals with systems acquisition. Objective 1 is broken down into the following subobjectives:

- By the end of 1994, institutionalize pollution prevention including hazardous materials minimization and management into the system acquisition process (concept, design, development, modification, maintenance and ultimate disposal) through the use of policies, procedures, training, contract provisions, and Federal Acquisition Regulation changes.

- Develop and incorporate procedures into system development milestone criteria that require: (a) Identification of hazardous materials, evaluation of environmentally acceptable alternatives, and selection of alternatives where indicated by life cycle analysis; (b) Identification of the remaining hazardous materials and the alternatives considered and reasons for their rejection; and (c) Estimates of the quantities of each hazardous material needed through the lifetime of the system; based on the most current concept of operations.

- Identify hazardous material use and waste generation for all acquisition programs, starting with C-17, B-2, T-l, HARM, TITAN-IV, Peacekeeper, JSTARS, and LANTIRN.

- Institute policies and procedures to reduce the use of ozone depleting chemicals 50% by 1995 from 1992 baseline, and eliminate the need to purchase ozone depleting chemicals by 1997.

- Replace hazardous material requirements in new system TOs, MIL-SPECS and MILSTDS with environmentally acceptable alternatives. Where none exist, prioritize the uses, select the ones with the highest potential improvement, and conduct a Science and Technology or Manufacturing Technology effort to develop alternatives.

- Identify material and process substitution needs critical to achieving pollution prevention objectives for integration into the Science and Technology Program.

- Obtain the resources required to accomplish the objectives.[3]

Programs that have been established to help attain the objectives regarding pollution prevention in weapon systems acquisition, as well as to minimize the adverse environmental impacts of new weapon systems, are described in the next section.

18.3   Environmental Issues in Weapon Systems Acquisition

In an effort to eliminate environmental problems associated with new weapon systems throughout the military, the Defense Department issued two documents in 1991 on environmental policy and procedures. DoDI 5000.2 and DoD Manual 5000.2-M require program managers to address environmental concerns at each step of the acquisition process. During the Concept Exploration and Definition Phase, the potential environmental effects of each alternative will be assessed and substantial effects will be noted.

ENVIRONMENTAL ANALYSIS

To comply with applicable environmental protection laws and regulations, an environmental analysis of new defense systems will begin at the earliest possible time. The initial environmental analysis will look at the entire life cycle of the program. Environmental effects will be identified in detail to allow integration with both economic and technical analyses steps that program managers must follow when acquiring new systems. It requires program managers to formally document potential environmental consequences with a more in depth Programmatic Environmental Assessment (PEA). The PEA is usually drafted during Phase I, Demonstration and Validation. It includes a description of the program; the alternatives to be studied; the potential environmental impacts; and the impact of alleviation procedures on schedule, cost, and site choices. Although security is still a program consideration, security classification does not free the program manager from the requirement to comply with all environmental laws or the requirement to integrate pollution prevention into technology, material, or process decisions.[8]

Proposed systems must be analyzed for their potential environmental impacts in accordance with Title 40, Code of Federal Regulations, Parts 1500-1508, "National Environmental Policy Act Regulations", and Executive Order 12114, "Environmental Effects Abroad of Major Federal Actions".

Program managers are now faced with the difficult task of balancing ever increasing environmental regulatory requirements with system performance, schedule and cost considerations. The constantly changing nature of government environmental policy dictates that program managers be cognizant of environmental laws. They must also anticipate the impact of new and pending environmental legislation on their acquisition strategy.

The policies and procedures governing environmental analysis within the Department of Defense are contained in Section 6-H of DoD Instruction 5000.2, "Defense Acquisition Management Policies and Procedures".

The initial assessment of environmental impacts should begin as soon as mission needs are determined and then be completed prior to Phase I, Demonstration and Validation.

The environmental analysis must look at the entire life cycle of the program. Environmental effects will be identified in adequate detail to allow integration with economic and technical analysis.

This environmental analysis contains a description of:

1. The program,

2. The alternatives,

3. Environmental impacts of each alternative throughout the system life cycle.

4. Potential for mitigation of adverse impacts, and

5. How the impacts and proposed mitigation would affect program performance, schedule, program cost and operations site locations.[9]

The PEA is a document of about 25-125 pages which identifies potential environmental consequences. The PEA is one of seven annexes to the Integrated Program Summary (IPS). The IPS is the main document used at each milestone review to decide whether to:
(a) Continue to the next phase of the acquisition process,
(b) Collect additional information, or
(c) Cancel the program.

HAZARDOUS MATERIALS IN SYSTEMS ACQUISITION

As was discussed in Section 18.2, the emphasis today is to shift away from simply managing hazardous waste. Program managers will now actively seek to reduce or avoid the use of hazardous materials in both processes and products. The ultimate goal is to prevent pollution by reducing generation of hazardous wastes to as near zero as possible.

In 1989, the Air Force created the Acquisition Management of Hazardous Materials (AMHM) program. This program was established in response to the growing economic and environmental costs of using hazardous materials and the recognition that the system acquisition process represents the first and most effective means to address the problem of hazardous materials.[10]

The goal of the AMHM program is to institutionalize consideration of hazardous materials issues in the weapon system acquisition process with an aim to minimizing hazardous materials use and hazardous waste generation throughout the weapon system life cycle.[11]

Where the use of hazardous materials cannot be reasonably avoided, there will be procedures for identifying, tracking, storing, handling, and disposing of the waste. DoD Directive 3150.2 and DoD Instruction 6050.5, "Hazardous Material Information System" provide guidance in this area.

The services are finding it difficult to replace many of their hazardous and environmentally unsafe materials. For example, the Navy uses radiators with chlorofluorocarbons (CFCs) to cool shipboard radars. Alternative cooling systems, that do not use CFCs, are both less efficient and much larger. Halon is another compound which is difficult to replace. It is used extensively by all services and is known for its ability to extinguish electrical fires without damaging electrical circuitry. Ironically, potential halon alternatives have proven to be even more hazardous.

In the future, program managers will need to maximize the use of environmentally safe materials and whenever possible use commercial items or existing defense systems that already have a proven environmental track record.

18.4   Air Force and Aerospace Industry Initiatives/Successes

In response to the growing concern over environmental issues, the Air Force and the majority of the aerospace companies have begun to implement a number of initiatives. These include programs to reduce the use of hazardous materials and prevent pollution.

The Air Force has established a number of very ambitious goals. These include:

1. To prevent future pollution by reducing the generation of hazardous waste and the use of hazardous materials to as near as zero as possible.

2. Reduce the use of the 17 EPA priority chemicals by 50 percent by the end of 1995.

3. Review all Air Force technical orders, military specifications and standards to eliminate all unnecessary use of the 17 priority chemicals.

4. Eliminate the purchase of all ozone depleting substances by the end of 1995.[12]

Successful hazardous materials management over the life cycle of a weapons system will require close coordination between the acquisition and logistic communities. The recent merger of Air Force Systems Command (AFSC) and the Air Force Logistics Command (AFLC) into the Air Force Materiel Command has facilitated this process. Since logistics operations generate 90 percent of the Air Force's hazardous waste, it is essential to establish a continual feedback mechanism between logisticians and weapon system designers.[13]

AEROSPACE INDUSTRY INITIATIVES

Many of the Aerospace Industries were early advocates of pollution prevention programs. General Dynamics (GD) has the Environmental Resource Management (ERM) program. This program focuses on the "efficient utilization of all materials and the elimination of hazardous materials generation, discharge, and disposal." Since 1984, through aggressive source reduction and recycling, GD has reduced hazardous waste 70 percent. This equates to the elimination of nearly 50 million pounds of hazardous waste. The program has realized a savings in materials and waste disposal costs of over $2 million per year.[14]

Boeing has a pollution prevention program called the Comprehensive Chemical Reduction program. Boeing's goal is to replace conventional processing materials with non-polluting, safer alternatives. Since 1984, more than 30 manufacturing process changes have been made to reduce the use of hazardous materials.[15]

The 3M Corporation started their Pollution Prevention Pays (3Ps) program in 1975. Their goal was to eliminate the dependence on hazardous materials rather than on treating pollution. By 1990, 3M cut their pollution per unit of output in half and saved in excess of $500 million.[16]

In a joint effort with the Air Force, Pratt and Whitney is using the Industrial Modernization Incentives Program (IMIP) to minimize hazardous waste. A $10 million investment seeks to:

1. Eliminate the use of hazardous materials in production and,

2. Reclaim wastes that are generated during the manufacturing process[17].

OPPORTUNITIES AND CHALLENGES

In light of diminishing waste disposal capacity and the extraordinary cost of waste cleanup, emphasis should focus on methods to minimize pollution at the source. In 1986 the Air Force's Scientific Advisory Board (SAB) stated that "the system acquisition process represents the first and most effective opportunity way to manage hazardous materials."[18]

Producing a million tons of hazardous waste each year, the Department of Defense is in a position to significantly improve the environment through an effective pollution prevention program. The Air Force has taken the lead among the services through initiatives on pollution prevention, personnel training, and on the reduction of hazardous materials/waste. A life cycle perspective of environmental issues provides valuable lessons for the design of future systems. The operation and maintenance of weapon systems generate 90% of all of the Air Force's hazardous waste. Feedback from logisticians in the design of future systems is critical.

Decisions made in the systems acquisition process have environmental effects throughout the life cycle. Often these ramifications are not recognized by either the acquisition personnel or by decision makers.

SYSTEMS LIFE CYCLE CONSIDERATIONS

One of the greatest opportunities is to address the use of hazardous materials early in the system acquisition process. Designing a system with an awareness of environmental issues will reduce the need for hazardous materials. This requires that the acquisition process must:

1. Identify and justify hazardous materials associated with each technology.

2. Conduct materials research to find substitute materials for hazardous materials.

3. Calculate life cycle environmental cost and make it visible for decision makers.

4. Examine environmental effects at each milestone review.

18.5   Conclusion

In the next decade, the Department of Defense will place increasing emphasis on environmental issues. The focus will be on three areas: pollution cleanup, compliance with environmental laws, and pollution prevention. In the area of pollution prevention, it is the acquisition process that presents the best opportunity for making significant progress. By incorporating environmental criteria in the selection of future weapon systems, tradeoffs can be made among materials and processes. Addressing pollution at the source is the best way to mitigate, or eliminate, future environmental problems.

References

[1] Cheney, Richard, "Environmental Management," Memorandum, 10 Oct. 1989.

[2] McPeak, Merrill A., "Environmental Leadership," Letter, 17 Apr. 1991.

[3] McPeak, Merrill A. and Donald B. Rice, "Air Force Pollution Prevention Program," Memorandum, 13 Nov. 1991.

[4] McPeak, Merrill A. and Donald B. Rice, "Air Force Pollution Action Plan," Memorandum, 7 January 1993.

[5] McPeak, Merrill A. and Donald B. Rice, "Air Force Ban on Purchases of Ozone Depleting Chemicals" Memorandum, 7 January 1993.

[6] Laskowski, Stanley L., "Pollution Prevention Act of 1990," Memorandum, 27 Nov. 1990.

[7] Air Force Policy Directive 19-4, "Pollution Prevention," Department of the Air Force, Headquarters U.S. Air Force, 1993.

[8] Williams, Roland, "Environmental Policy," Program Manager, May-June 1989, p.7.

[9] Department of Defense, DoDI 5000.2, "Defense Management Policies and Procedures," 15 September 1990, Part 4, Section F, annex E, p.G-I-5.

[10] "Acquisition Management of Hazardous Materials Technical Integration Plan," MITRE, 7 Aug. 1991, p. xi.

[11] "Acquisition Management of Hazardous Materials Identification and Evaluation Process," MITRE, Apr. 1991, p. xi.

[12] "Pollution Prevention and the Acquisition of Aircraft Weapon Systems," MITRE, Jun. 1992, p. 3-7.

[13] Ibid, p. 3-8
[14] Ibid, p. 3-4.
[15] Ibid, p. 3-2.
[16] Ibid, p. 3-1.
[17] Ibid, p. 3-3.
[18] Ibid, p. 1-2.

Chapter 19

INTERNATIONAL PROGRAMS

Craig M. Brandt and Lt. Col. Charles M. Farr
Air Force Institute of Technology

19.1 Introduction[1,2]

Since World War II, the United States has used the transfer of military systems and equipment to allies and friends as an integral part of a foreign policy based on mutual security. In the earliest years, nearly all the material which was sent abroad was surplus to American needs and provided at no cost to the recipient. This grant aid program became known as the Military Assistance Program and represented U.S.-funded gifts to foreign militaries. By the early 1960s, the economies of many of our allies, especially in Europe, had recovered from the war, and these nations were able to share in the defense burdens. The countries could purchase weapons from the U.S. rather than relying on donations of American surplus.

The global military commitments of the U.S. were accompanied by an expansion of the countries with which we had arms transfer relationships, and this growing business of supplying our allies became important not only in foreign policy terms but as a segment of military logistics support. Sales as a percentage of total transfers continued to rise, and by 1992, although there were still some significant grant programs, foreign military sales was by far the most common means of transfer. By the early 1990s, annual sales of defense materials of all types had reached a level of $15 billion and in many cases the number of foreign equipments supported by the logistics establishment was greater than that in the U.S. inventory.

Foreign militaries are by no means totally reliant on U.S. weapons. While in the immediate post-WW II years the U.S. was the principal supplier of weapons for the West, over the years the technological and manufacturing bases of many other nations has enabled many of our allies to produce first-rate weapon systems. By the mid-1970s, it became the U.S. policy to look at possible cooperative ventures in producing or manufacturing weapons in order to create standard equipments which would enhance war fighting capability among our alliances, reduce total costs among partners by sharing costs and risks, and balancing international trade in armaments. Originally, cooperation was simply joint production of weapons previously developed by a single nation, but today cooperation has expanded to include research and development, comparative testing of weapons, and industrial cooperation as well.

Armaments cooperation is not totally separate from foreign military sales since in many instances manufacturers selling major weapons agree to offsets, that is, a reciprocal arrangement in which some form of cooperation may be offered as part of the terms of sale.

Today statutes and regulations have created a separate set of guidelines for the conduct of foreign military sales and for cooperative military programs. These additional requirements on top of the normal acquisition and logistics systems add an element of difficulty to the management of these international logistics programs, but international involvement in both the acquisition of major systems as well as the providing of follow-on support show no signs of diminishing.

19.2   Foreign Military Sales

As an element of foreign policy, the sale of weapons to foreign countries falls under the general control of the Department of State which must determine if a country should be allowed to receive American arms and to what extent. The role of the Department of Defense is to implement the sale, once it has been approved within the context of overall foreign policy. In general, the sale of defense articles is based on two relationships, the first between the United States and the purchasing country, and the second between the military department and the supplier of the material desired. Two basic laws control our international arms transfer programs, which fall under the general rubric of "security assistance": the Foreign Assistance Act of 1961 and the Arms Export Control Act of 1976. As amended annually, these statutes provide the legal basis for the conduct of the export of weapons.

For the most part, the acquisition and logistics involved in foreign sales are exactly the same as those used to support U.S. forces.[3,4,5,6] Because of the specialized nature of the sales business, there are a few organizations which are superimposed on the USAF organization to carry out the security assistance programs. The Deputy Under Secretary for International Affairs (SAF/IA) has the primary responsibility for the central management, direction, guidance and supervision of the Air Force portion of international programs. The Assistant Secretary of the Air Force (SAF/AQ) coordinates in the development of offers of major weapon systems and oversees their procurement. The Air Force Security Assistance Center (AFSAC) oversees the performance of Air Force Materiel Command organizations in the support of international customers. The Air Force Security Assistance Center (AFSAC), under the guidance of the Air Force Materiel Command International Affairs office (AFMC/IA), oversees the performance of AFMC organizations in the support of international customers. The Air Force Security Assistance Training (AFSAT) is responsible for the coordination of training of international students both in USAF schools and in country.

THE SALES PROCESS

The sales process starts with the receipt from a foreign country of a letter of request for the materiel or services desired. While headlines in the press frequently stress sales of major aircraft or missiles, in fact countries may request virtually any item or service which is used in the U.S. military. For example, sales may include aircraft, spare parts, training, publications, ammunition, modifications, medical supplies, maintenance support, or contractor assistance. Although all requests must be approved by the Department of State for consistency with U.S. foreign policy interests, the requests ultimately are acted on in the military departments. In the Air Force, three offices receive the letters of request: SAF/IA for major system sales, AFSAC for follow-on logistics support, and AFSAT for training requirements.

Once a purchaser's request has been validated for completeness, eligibility, and releasability, a Letter of Offer and Acceptance (LOA) may be prepared by the appropriate service activity. Major system sales LOAs are prepared by SAF/IA, logistics support LOAs by AFSAC, and training LOAs by AFSAT. The LOA contains estimated costs obtained from Systems Program Offices, Air Logistics Centers, etc., which may be based on contractors' quotes or projected costs and a forecast of item availability. It is DoD policy to ensure that on a major system sale all possible logistics and training requirements have been considered. This is done in order to avoid selling a systems only to discover that there are no spares, technical publications, or training available. The LOA contains not only the items to be sold, but a variety of financial and logistics codes which provide billing, shipping, and transportation information.

After the final review of the offer at both the Department of Defense and State levels, the offer is sent to the purchasing government which normally has 60 days to accept the LOA. Upon acceptance, this document becomes the contract between the two countries for the material and services offered.

The agreement must then be implemented by the military department to deliver the required articles.

After the initial deposit has been received, the Defense Finance and Accounting Service (DFAS-DE/I) may issue the obligational authority to the military department and an implementing directive can be prepared advising various organizations of their responsibilities in carrying out the sale. Price adjustments or material changes to the program may be accomplished by processing amendments and/or modifications throughout the life of the case.[7]

In supplying materiel to the foreign customer, the normal logistics system of the military department responds in much the same manner as for U.S. military requirements. If regulations permit, material may be issued from existing DOD stocks; otherwise it must be purchased from a vendor. Under ordinary circumstances, material is not shipped directly to the purchasing country but instead is directed to a freight forwarder in the United States which acts as shipping agent for the country. The freight forwarder in turn arranges onward transportation to the purchaser. Thus, the freight forwarder is an essential link between the American logistics system and that of the purchasing country.

The norm in transportation is to rely on private the transportation industry to the extent possible for the movement of Foreign Military Sales (FMS) material. In this case, the country pays the transportation bill largely outside of the FMS payment system. In other cases, however, the Defense Transportation System (DTS) may be used. A country may negotiate use of DTS even for delivery of materiel in-country. In all cases where DTS is employed, the country is charged for these services in accordance with standard procedures.

FOLLOW-ON LOGISTICS SUPPORT

While the procedures of letter of request and letter of offer are acceptable for major end items, they are exceedingly cumbersome for repetitive buys of items such as spare parts which constitute the bulk of actions for follow-on logistical support. Consequently, there are specialized mechanisms to facilitate the ordering of items after the initial purchase. The first of these is a blanket order case, which establishes a dollar-value case for a category of commodity, such as aircraft spares or technical manuals. This case does not further define the items which may be ordered, and items may be ordered within a dollar ceiling as long as funds are available. This permits a country to negotiate a sales case only once, and then allows individual items to be ordered in a streamlined manner. For example, items of supply are requisitioned using standard Military Requisition and Issue Procedures

(MILSTRIP). Typically, blanket order cases are used for spare and re-pair parts, publications, minor modifications, technical assistance services, training, and reparables.

The blanket order method for material allows a foreign country access to U.S. stocks to the extent that inventory levels are above the reorder point. If the inventory balance is below the reorder point, foreign requisitions must be back-ordered or purchased separately. Consequently, on the average, the supply response time to FMS requisitions of this type is slower than U.S. forces experience. Prices might also be elevated if items are purchased expressly for the foreign customer, and there is no attempt to take advantage of quantity discounts.

To avoid these difficulties for spare and repair parts, the Cooperative Logistics Supply Support Arrangement (CLSSA) has been developed. Under this arrangement, the country first invests money in the military department supply system. The amount is based on the usage the country anticipates based on its equipment population and operational schedules. This money then is spent by wholesale item managers to augment U.S. stocks to accommodate the larger number of equipments being supported. This investment allows the country to become a partner in the U.S. supply system and results in the country's access to material below the conventional reorder point. This investment is handled through a special FMS case, the Foreign Military Sales Order (FMSO) I.

After the augmentation of material in the U.S. system is complete, the country may then place orders through a Foreign Military Sales Order II requisition case, which is like a blanket order case in that a dollar ceiling is specified, a time period is set, but no items are specifically identified.

Although the initial acquisition of weapon systems under FMS procedures is an important step, both in political and military terms, the follow-on logistics represents the bulk of the day-to- day work in international sales. Approximately two-thirds of the sales value is for logistics support.

FINANCIAL PROVISIONS

Because of the requirements of the Arms Export Control Act (AECA), there are some additional limitations placed on standard military procedures, especially in the financial area. The AECA requires that the U.S. government operate the sales system at no profit and no loss. That is, the American taxpayer cannot subsidize sales abroad, nor can the government use sales as a profit-generating venture. This has two immediate implications for the foreign buyer with respect to the LOA. First, the prices included in the LOA are estimates, but acceptance of the offer means that the buyer will pay for the item, regardless of its final cost. The terms and conditions of the LOA explain this clearly, the result being that the LOA becomes in fact a cost-reimbursement contract. Without this provision, the U.S. would have to appropriate funds to accommodate price increases. Of course, if the costs are lower than the estimates, the purchaser will be liable only for the lower amount.

Second, the financial system employed in foreign military sales requires that the customer pay in advance for all purchases. Consequently, the acceptance of the LOA must include an initial deposit that has been calculated to produce the funds which will be used on the foreign country's behalf in the first six months of a program. An estimated payment schedule is included with the offer and shows when monies will be due in the U.S. prior to their anticipated expenditure. This schedule covers cost of materiel, progress payments, contractor holdback, termination liability, and special FMS administrative and accessorial charges. Quarterly bills are sent to the customer, indicating the materiel and services delivered and requesting the money to be paid in accordance with this schedule.

To pay for the U.S. government resources used managing a foreign sale, a three percent administrative charge is added to the material cost to cover the costs of sales negotiations, case implementation, program control, computer programming, accounting, and other administrative activities. In order to recoup previous government investment, foreign purchasers are charged a pro rata share of nonrecurring research, development, test and evaluation as well as nonrecurring production costs. For items procured from a vendor, a charge is levied to cover costs of contract administration and quality assurance. In essence, all government overhead functions are reimbursed by the purchaser under the terms of the normal foreign military sale. For those cases where management of a sale exceeds normal administrative functions, the efforts of the U.S. government employees may be charged directly to the customer as a part of the sale.

CONTRACT TERMS

The terms and conditions of the LOA state that DoD will procure defense articles and services consistent with DoD regulations and procedures, thus affording the purchaser the same protection and benefits that apply to defense procurement. However, the prices paid by a foreign customer may differ from those paid by DoD since the Defense Federal Acquisition Regulation Supplement (DFARS) Section 225.73 recognizes that costs of doing business with a foreign government differ from those of dealing with the United States. Such differences as selling expenses, configuration studies, product and post-delivery support may exist and are legally included in a contractor's price. On an exceptional basis, the foreign purchaser may submit a request for a specific vendor to be the sole source of supply for an item. If the request is valid and meets DoD justification criteria, it will be honored and the identified vendor will used.[8]

The general pricing rules for the sale of materiel and services insure that there is no U.S. subsidy. For items from a vendor, all applicable costs must be recovered and for items which are sold from stock, the actual value or replacement value must be charged, depending on the circumstances of the sale.

19.3   International Defense Cooperation

In the environment of foreign military sales, U.S. allies are normally treated as customers who are purchasing defense equipment and services that have been developed and/or produced by the United States. As mentioned earlier, however, another important mechanism has emerged in the world of international armaments. Since the mid-1970s, countries have increasingly entered into cooperative projects that differ significantly from the sales arena. In these cooperative arrangements, the participating countries are partners who share costs, risks, management responsibilities, technology, design/development tasks, and production in accordance with a negotiated agreement known as a Memorandum of Understanding (MOU).[9]

THE EMERGENCE OF COOPERATIVE PROJECTS

U.S. participation in cooperative projects has grown over 3000 percent since World War II. While most of the growth has occurred since the mid-1970s, it has come with considerable controversy. Advocates of cooperation point to numerous potential advantages including the ability to share the rapidly increasing costs of high technology weapons. The enormous risks associated with some of today's programs prompted a Boeing executive to describe "going it alone" as tantamount to a "bet the company" proposition. Other potential advantages of cooperation include: access to global markets that would otherwise become increasingly closed to U.S. companies, access to increasingly sophisticated technology that has surpassed what is available in the United States in some areas, the opportunity to reduce the development time for new weapons by taking advantage of work already done in other countries, the opportunity to also reduce the cost of development by eliminating redundant investments in the same technologies among allied countries, the development of more standardized systems that create military advantages both in the operation of the systems in battle and in the logistics of supporting those systems when they are deployed around the world, and the increased political cooperation that may accompany these shared investments.

Opponents of cooperation have voiced their concerns about the potential downside of these joint projects. The two most frequently heard arguments involve technology and jobs. Some people worry that the U.S. may give away more technology than it receives, and that this potential loss of technology may hurt U.S. industry as well as prompting national security concerns. There are related economic concerns that focus primarily on the notion that cooperative programs cause U.S. jobs to be exported abroad. The economic concerns are also extended to include the fear of creating competitors who may eventually take business away from U.S. companies. There are also concerns that these project types require complex organizational structures that may impair decision making, and that demands for the distribution of work among the various participating industries may lead to inefficient and costly production programs.

Despite the on-going controversy and the difficulty of managing these complex ventures, the U.S. continues to participate in them. Many experts now believe that the economies of individual nations have in fact become globally interdependent. In this view, the world is no longer dominated by the post-World War II economic and technological strength of the United States. While still a world leader, the U.S. now faces technologically sophisticated allies who are able to demand a greater role in the development and production of weapons. So, while the foreign military sales program continues, we have a new array of international project types to contend with as well.

INTERNATIONAL COOPERATIVE ACTIVITIES

International cooperation can occur across the entire life cycle of the military weapons acquisition process. Early in the process, countries may exchange information about perceived security threats and identify any common interests in working together. The NATO alliance has a well organized forum for this kind of interaction with a variety of regularly scheduled meetings, senior national representative meetings, bilateral staff talks, and so on. Many countries also exchange technical information and work to identify common interests in developing new technology. Data Exchange Agreements (DEA) and Information Exchange Programs (IEP) are frequently used to facilitate this process. DEAs and IEPs are written agreements, usually between two nations, that specify particular areas of technology in which the nations intend to share scientific information. These agreements have often been the genesis for subsequent joint weapons programs. Another forum for exchange of scientific information is the Engineering and

Science Exchange Program (ESEP) in which nations host visiting scientists in their laboratories.[10]

There are several types of cooperative projects that are used during the development phases of the acquisition process, i.e. concept definition, demonstration and validation, and engineering and manufacturing development). Joint test programs are one such project type. The Foreign Comparative Test program is managed by the Director of Test and Evaluation within the Office of the Under Secretary of Defense for Acquisition. The objective of the program is to reduce duplication in R&D and accelerate the fielding of new or modified weapons by taking advantage of technologies developed in other countries. The program also seeks to stimulate competition and the strengthen the technology base of all participating countries by facilitating technology transfer. Further, the program encourages industrial teaming arrangements.[11]

Potential items to be evaluated are nominated by the military services (Air Force/Army/Navy/Marines). The nominated projects are then reviewed and selected by a committee from the Office of the Secretary of Defense (OSD). The selection criteria favor programs that meet documented U.S. requirements and in which the allied system is well proven (preferably in production and/or operational use). The program should also offer a performance, cost, or schedule advantage over any U.S. system currently in use or under development. There have been a total of 273 projects from 1980 through 1992. Approximately twenty percent of the projects have resulted in U.S. military procurement of the system. Recent successes have included a British aerial target vectoring system, German traveling wave tubes, French spot satellite digital imagery, Swedish remote controlled minesweeping system, and an Israeli stand-off air launched guided weapon.

The other major category of development projects is referred to as codevelopment. Codevelopment normally occurs in one of two ways, either as a single project or under a concept known as Family of Weapons. The vast majority of codevelopment work occurs in the single project category in which the participating countries share the development of one particular system. If the project is large enough and the countries agree, a jointly manned project office is established in one of the countries. In this way, all countries have managers participating in the daily work and decision making on the program. These representatives are not merely liaison personnel; they are experienced managers with project responsibilities in areas such as engineering, manufacturing, and logistics. Recall that the rights and responsibilities of all the participating nations in such a project are delineated in the MOU which is the result of a negotiation process. The MOU should address important issues such as technology, future sales of the system to countries not in the partnership, the process for making decisions and for resolving conflicts, the distribution and sharing of work among the countries, the sharing of costs, the conditions and timing under which countries are permitted to withdraw from the project, intellectual property rights, and what legal and regulatory guidelines will be used.

The other category of codevelopment, Family of Weapons programs, has not been used very much. The concept envisions one or more countries developing a system while another team develops a related capability. For example, the United States has developed an advanced medium range air to air missile (known as AMRAAM) while several European countries were to develop a short range version of the same weapon (ASRAAM). Under this concept, each country would only need to fund the development of one of the weapons. In this example, however, the European development program lagged far behind the U.S. AMRAAM and it does not appear that the European system will be forthcoming. The intent was that countries participating in the arrangement might coproduce or purchase a system that they did not have to develop first. Again, for some reason the concept has not been used much in practice.

Finally, joint military production programs are generally referred to as coproduction. There are two primary methods that have been used for coproduction programs, licensed production and production sharing arrangements. Under licensing arrangements, technology and manufacturing capability is transferred to the receiving country which then produces units and pays licensing fees or royalties to the company that originally developed the technology. For instance, Japan has historically produced such U.S. weapons as the F-104 and F-15 fighters under licensing agreements. Licensing arrangements are usually for quantities to be used only in the country of the licensee and are not intended for export sales to other countries. Production sharing arrangements are more like commercial joint ventures. The partners share production work in some agreed fashion. The partners might split up the system so that each contributes what they are best at doing. In some cases, however, there may be more than one final assembly line and the partners each produce complete systems. Under most production sharing arrangements there is also an expectation of export sales to countries outside the partnership.

Two other observations about joint production programs should be mentioned. If the partners also had a codevelopment program that preceded production, the MOU will generally be updated or renegotiated for production. Also, there is a hybrid production arrangement that is occasionally used. This arrangement provides for limited coproduction or coassembly in what is otherwise a foreign military sales program. These programs usually result from political pressure on the United States to grant some work on a military program in the purchasing country. An example is Egyptian coproduction on the F-16 program.

POTENTIAL IMPACT ON PROJECT PERFORMANCE[12,13]

U.S. participation in cooperative projects can potentially result in positive or negative impacts on the traditional measures of performance for DoD projects (cost, schedule, and technical performance). With respect to the schedule, successful joint testing or family of weapons programs may actually reduce the time it takes to field a weapon in support of a new requirement. However, typical codevelopment or coproduction programs may cause the time to field a weapon to increase. A major reason for this adverse impact is that the process for negotiating and agreeing on an MOU typically takes one or two years. There may be other occasions during a cooperative project where gaining a consensus on a difficult issue may take extra time. With the exception of joint testing, most experts agree that cooperative programs take longer than U.S.-only projects.

The total combined cost to all countries in a cooperative project is frequently greater than the cost might have been to a single country doing the program on their own. However, two important points must also be considered. First, the actual cost to participants in a cooperative venture is limited to a percentage of the total cost and this percentage is established at the outset. Therefore, successful joint projects usually represent a cost savings to participating countries. Second, sometimes a country cannot accomplish a program on their own because of technological or funding limitations. Partnerships become the only means to retain indigenous capabilities at an affordable cost. Countries do, however, accept the risk that other participants may withdraw from the program and consequently change the cost sharing equation.

The technical performance criterion presents mixed results, particularly to the United States. Partnerships frequently produce some negotiation and inevitable compromises on the performance requirements for the weapon system. There is frequently pressure exerted on the United States to reduce requirements because other countries do not need and cannot afford the complete set of requirements levied by the U.S. military. This situation occurs because the U.S. defense budget and acquisition process is ten times larger than that of France, Germany, the U.K., or Japan. Further, the U.S. policy to accept worldwide security responsibilities imposes a more stringent array of requirements for our weapons than is typical in other countries. On the positive side, the U.S. has gained access to world-class technology in a number of areas, particularly during the last decade. Also, one might consider that the restraint of our partners may moderate the U.S.

tendency to pursue the "last ten percent of performance at a substantial cost penalty."

The net result of cooperative projects is probably that in most instances they take longer but save money. Technologically, they produce conflicts about performance requirements but allow access to superior technology elsewhere in the world.

THE FUTURE OF ARMAMENTS COOPERATION

Although cooperative projects have produced considerable debate and controversy, their potential advantages continue to attract the attention of government and industry managers. Further, there is widespread evidence that the economies and industries of individual nations have indeed become globally interdependent. There has been steadily increasing empirical evidence of this globalization as more and more companies formed joint ventures and strategic alliances with selected partners in the United States, European Community, and Pacific Rim. Additionally, in its June 1992 report to Congress, the Office of Technology Assessment noted that the U.S. must consider the strategy of globalization in its efforts to maintain the defense technology and industrial base in the face of further downsizing and consolidation.

There are two obvious defense markets that the U.S. must pay attention to, Europe and the Pacific Rim. Members of the European Community have indicated that they will increasingly investigate "Europe-only" possibilities on new armaments programs. However, they have also indicated that significant U.S. involvement is expected for three reasons. First, the size of the U.S. market cannot be ignored. Second, the U.S. still has much of the leading technology in defense applications and other countries will want access to it. Finally, the U.S. has significant experience in integrating large systems, a capability that is lacking or underdeveloped in many countries.

As for the Pacific Rim, major growth is expected during the 1990s and much of it is already underway. There is presently no strong regional direction such as that provided by the European Community, and some Pacific Rim countries are more interested in cooperative defense projects than others. The complete technological and economic dominance of Japan is likely to be reduced. It is likely that the U.S. will pay more attention to the prospects for defense cooperation in this part of the world. Many European countries are already doing so.

Finally, what about the trends in the overall use of these project types? Will their use increase or will they die in the face of a new wave of protectionism? European and U.S. experts have suggested that the following is likely to happen. As downsizing really begins to affect defense employment, there will probably be some predictable protectionist responses. However, in the long term, the logic for sharing costs and risks will be more compelling than ever before. As the Office of Technology Assessment (OTA) report suggests, globalization will probably be part of our industrial base strategy.[14]

References

[1] Brandt, Craig M., ed., "Military Assistance and Foreign Policy," Air Force Institute of Technology, Wright-Patterson AFB, OH, 1990.

[2] Defense Institute of Security Assistance Management, "The Management of Security Assistance, Defense Institute of Security Assistance Management, Wright-Patterson AFB, OH, 1992.

[3] Department of the Air Force, Regulation AFR 130-1, Security Assistance Management.

[4] Department of the Army, Regulation AR 12-8, Foreign Military Sales Operations/Procedures.

[5] Department of Defense, DoD Manual 5105.38-M, Security Assistance Management Manual.

[6] Department of the Navy, NAVSUP Publication 541, Security Assistance Manual.

[7] Department of Defense, DoD Manual 7290.3-M, Foreign Military Sales Financial Management Manual.

[8] Department of Defense, Defense Federal Acquisition Regulation Supplement, Subsection 225.73.

[9] Defense Systems Management College, Guide for the Management of Multinational Programs, 2nd edition, Fort Belvoir, VA, 1987.

[10] Department of Defense, "International Cooperative Programs Management Manual," Draft DoD 5134.1-M, Office of the Under Secretary of Defense (Acquisition), April 1991.

[11] Department of Defense, "Foreign Comparative Testing Program," Handout, Manager, Foreign Comparative Test Program, Office of the Under Secretary of Defense for Acquisition, Director of Test and Evaluation.

[12] Farr, Charles Michael, "An Investigation of Issues Related to Success or Failure in the Management of International Cooperative Projects, Dissertation, University of North Carolina, Chapel Hill, NC, 1985.

[13] Farr, C. Michael, "Managing International Cooperative Projects: Rx for Success," Chapter in Global Arms Production: Policy Dilemmas for the 1990s, Center for International Affairs, Harvard University, 1992.

[14] Congress of the United States, Office of Technology Assessment. "Building Future Security: Strategies for Restructuring the Defense Technology and Industrial Base," Washington, DC, Government Printing Office, June 1992.

Chapter 20

ACQUISITION EDUCATION

20.1 Introduction

Following the enactment of the National Defense Authorization Act for Fiscal Year 1991 and the Defense Acquisition Workforce Improvement Act, the Department of Defense created an acquisition university as a consortium of 15 Army, Navy, Air Force, and Defense Logistics Agency schools and activities. The consortium component activities are coordinated by the university president with a small administrative staff placed under the Deputy Under Secretary of Defense for Acquisition. The DAU, as consortium of defense schools, coordinates and tailors education to the needs of career personnel serving in DoD acquisition positions. It serves as the DoD center for the development of education, training, research, and publication capabilities in the area of acquisition.[1]

20.2 Defense Acquisition University Structure and Operations

The consortium structure builds upon the strengths of existing schools and provides the flexibility to include other institutions to meet new or specialized needs of career acquisition professionals. Each DAU member school retains its own command structure. Its relationship to DAU is spelled out in a Memorandum of Agreement (MOA). A council composed of senior DoD officials and chaired by the Under Secretary of Defense for Acquisition (USD(A)) provides university policy and operations oversight.

Fostering debate, examining acquisition issues and developing university faculty by supporting research, symposia, conferences and maintaining publication capabilities assures excellence in acquisition management education. These endeavors keep the DAU in the forefront of acquisition issues. Topics are chosen to apply professional acquisition expertise to specific needs of the DoD community.

Table 20.1: Acquisition Functions and Career Fields.

ACQUISITION FUNCTIONS	CAREER FIELDS
• Acquisition Management	• Program Management • Communication-Computer Systems
• Procurement and Contracting	• Contracting • Purchasing • Industrial Property Management

• Systems Planning, Research,
  Development, Engineering,
  and Testing
	• System Planning, Research,
	  Development, and Engineering
	• Test and Evaluation Engineering

• Production	• Manufacturing and Production • Quality Assurance
• Acquisition Logistics	• Acquisition Logistics

• Business, Cost Estimating,
  and Financial Management
	• Business, Cost Estimating,
	  and Financial Management

• Auditing	• Auditing

DAU works in conjunction with functional boards to identify competencies and translate needs to courses. Functional boards are established for each of the seven functional acquisition areas (see Table 20.1) to provide oversight of management and career program execution. These boards consists of senior level acquisition of the DoD components who advise the USD(A) on issues of career development and recommend mandatory training for their functional area including required curriculum/course content.

DAU manages training resources and administers the acquisition education and training program to prepare professionals for effective service in the DoD acquisition work force. This includes developing, submitting and defending funding through the Planning, Programming and Budgeting System (PPBS). Funds are suballocated for course development, presentation and student attendance. DAU tracks budget execution and analyzes use of funds to assure optimum return on investment.

The DAU is the executive agent for managing and coordinating the delivery of mandatory acquisition courses. This function was formerly known as the Acquisition Enhancement (ACE) program.

DAU courses are offered at the basic, intermediate, and senior levels for DoD acquisition managers. Offerings range from basic level training on Management of Defense Acquisition Contracts, provided by several DAU schools, to the 20-week Program Management Course provided at the Defense Systems Management College (DSMC) and the Senior Acquisition

Course taught at the Industrial College of the Armed Forces (ICAF). The DAU focuses on career development needed for certification and membership in the Defense Acquisition Corps. Course lengths range from a few days to several weeks and are provided at sites worldwide, throughout the year, to meet the training requirements of the Defense acquisition community. In addition to providing resident instruction and instructors to travel to local facilities, the university provides innovative learning opportunities, including satellite instruction, video tutorials, equivalency examinations and correspondence courses.

The DAU publishes an annual DoD Acquisition Education and Training Catalog and a schedule of course offerings. The catalog describes mandatory courses for selected career functional areas in addition to requirements for advancement in specific fields. Courses are grouped by career field and career levels. The catalog also provides information on consortium members and points of contact for course quota management. Course descriptions, prerequisites and course sponsors are addressed individually and grouped by date and location.

The following are members of the DAU consortium and provide acquisition courses:

1. Air Force Institute of Technology (AFIT).

2. Army Logistics Management College (ALMC).

3. Army Management Engineering College (AMEC).

4. Defense Contract Audit Institute (DCAI).

5. Defense Logistics Civilian Personnel Service Support Office (DCPSO).

6. Defense Systems Management College (DSMC).

7. European Command (EUCOM) Contracting Training Office (CTO).

8. Industrial College of the Armed Forces (ICAF).

9. Information Resources Management College (IRMC).

10. Lowry Technical Training Center (LTC).

11. Naval Facilities Contracts Training Center (NFCTC).

12. Naval Supply Systems Command (NAVSUP) Contract Management Directorate.

13. Naval Warfare Assessment Center (NWAC).

14. Navy Acquisition Management Training Office (NAMTO).

15. Office of the Assistant Secretary of the Navy (OASN(RD&A)).

DAU provides centralized control for attendance at their courses through the Army Training Requirements and Resources System (ATRRS). Allocation of DAU course quotas, registration and course completion information is accomplished electronically. Course quota requirements are developed and allocated annually. Quota refinement and course adjustments assure optimum use of DAU training capacity. DAU coordinates with the DoD Directors of Acquisition Career Management (DACMs) to make adjustments as necessary.

20.3   New Requirements for Education and Training

The Defense Acquisition Workforce Improvement Act expanded the number of acquisition functions to be addressed by DoD's acquisition education and training program. Specifically DoD establishes education, training and experience requirements for the following twelve career fields:

1. Program Management.

2. Communications/Computer Systems.

3. Contracting.

4. Purchasing.

5. Industrial Property Management.

6. Systems Planning, Research, Development and Engineering.

7. Test and Evaluation Engineering.

8. Manufacturing and Production.

9. Quality Assurance.

10. Acquisition Logistics.

11. Business, Cost Estimating, and Financial Management.

12. Auditing.

These career fields are grouped into seven acquisition functions as shown in Table 20.1.

For purposes of developing career ladders and an educational framework to support the education, training and career development needs of the acquisition workforce, the above acquisition positions have been grouped into seven principal acquisition functions as shown in Table 20.1.

DoD Manual 5000.52M, Career Development Program for Acquisition Personnel, provides instructions for career development of persons serving in the acquisition workforce. Education, training, and experience experience requirements are specified at the entry, intermediate, and senior levels for those position categories added by the statute. This manual establishes the framework for the educational developments of personnel in acquisition positions, from the basic to senior levels, required by the statute. The senior course is the "capstone" of this framework.

DAU courses are grouped by career field and divided into three career levels. While generally associated with a specific career field, and career level, some courses are multi-functionally related and required at various career levels. The basic level, *Level I*, courses are designed to provide fundamental knowledge and establish primary qualification and expertise in the individual's career fields, job series, or functional area. At the intermediate level, *Level II*, specialization is emphasized. The courses at this level are designed to enhance the acquisition personnel capabilities in the individual's primary specialty or functional area. At the senior level, *Level III*, acquisition training emphasizes management of the acquisition process and the latest methods being implemented in the career field or functional area. Each of these levels typically corresponds to particular GS/GM levels or military ranks which have been defined by the individual military department or agency; however, grade is not generally a requirement for course enrollment.

References

[1] Office of the Under Secretary of Defense (Acquisition), "Defense Acquisition University Catalog, 1992-1993," Vol. 1, ADS-92-01-CG, Defense Acquisition University, Alexandria, VA, 1992.

GLOSSARY OF ABBREVIATIONS AND ACRONYMS

ABDR	Aircraft Battle Damage Repair
ACAT	Acquisition Category
ACC	Air Combat Command
ACE	Acquisition Enhancement (Program)
ACI	Analytical Condition Inspection
ACO	Administrative Contracting Officer
ACSN	Advanced Change/Study Notices
ACWP	Actual Cost of Work Performed
ADM	Acquisition Decision Memorandum
ADPE	Automatic Data Processing Equipment
ADTC	Armament Development and Test Center
AECA	Arms Export Control Act
AEDC	Arnold Engineering and Development Center
AF	Air Force
AFAE	Air Force Acquisition Executive
AFBCA	Armed Forces Board of Contract Appeals
AFCAA	Air Force Cost Accounting Agency
AFCC	Air Force Communications Command
AFCESA	Air Force Civil Engineering Support Agency
AFDTC	Air Force Development Test Center
AFFTC	Air Force Flight Test Center
AFIT	Air Force Institute of Technology
AFLC	Air Force Logistics Command
AFMC	Air Force Materiel Command
AFOSR	Air Force Office of Scientific Research
AFOTEC	Air Force Operational Test and Evaluation Center
AFPRO	Air Force Plant Representative Office
AFR	Air Force Regulation
AFSAC	Air Force Security Assistance Center

AFSARC	Air Force Systems Acquisition Review Council
AFSAT	Air Force Security Assistance Training
AFSC	Air Force Systems Command
AFSC	Air Force Specialty Code
AFSOC	Air Force Special Operations Command
AFWCCS	Air Force Wing Command Control System
AGM	Air-to-Ground Missile
AGMC	Aerospace Guidance and Metrology Center
AGS	Aircraft Generation Squadron
AIAA	American Institute of Aeronautics and Astronautics
AIM	Air Intercept Missile
ALC	Air Logistics Center
ALMC	Army Logistics Management College
AMEC	Army Management Engineering College
AMRAAM	Advanced Medium-Range Air-to-Air Missile
ALU	Arithmetic/Logic Unit
AMARC	Aerospace Maintenance and Regeneration Center
AMC	Air Mobility Command
AMHM	Acquisition Management of Hazardous Materials
AMRAAM	Advanced Medium Range Air to Air Missile
AMS	Avionics Maintenance Squadron
AMSDL	Acquisition Management Systems & Data Requirements Control List
AMU	Aircraft Maintenance Unit
ANSI	American National Standards Institute
APB	Acquisition Program Baseline
APB	Amended President's Budget
APTS	Acquisition Program Tracking System
ASC	Aeronautical Systems Center
ASRAAM	Advanced Short Range Air to Air Missile
ASP	Acquisition Strategy Panel
ASQC	American Society for Quality Control
ASR	Alternative System Review
ATC	Air Training Command
ATD	Advanced Technology Demonstration
ATE	Automatic Test Equipment
ATTD	Advanced Technology Transition Demonstration
AU	Air University
AWACS	Airborne Warning and Control System
BAFO	Best and Final Offer
BCWP	Budgeted Cost for Work Performed
BCWS	Budgeted Cost for Work Scheduled

BES	Budget Estimate Submission
BMD	Ballistic Missile Defense
BMO	Ballistic Missile Office
BPPBS	Biennial Planning, Programming, and Budgeting System
C/SCSC	Cost/Schedule Control Systems Criteria
C/SSR	Cost/Schedule Status Reviews
CAE	Component Acquisition Executive
CAID	Clear Accountability In Design
CALS	Computer-aided Acquisition and Logistics Support
CAO	Contract Administration Office
CASC	Cataloging and Standardization Center
CBD	Commerce Business Daily
CCB	Configuration Control Board
CCBD	Configuration Control Board Directive
CCP	Contract Change Proposal
CDR	Critical Design Review
CDRL	Contract Data Requirements List
CE	Concept Exploration and Definition
CFSR	Contract Funds Status Reports
CI	Configuration Item
CICA	Competition in Contracting Act
CINC	Commander-in-Chief of the Unified or Specified Command
CLIN	Contract Line Item Number
CLSS	Combat Logistics Support Squadron
CLSSA	Cooperative Logistics Supply Support Arrangements
CM	Configuration Management
CMU	Cheyenne Mountain Upgrade (Program)
CO	Contracting Officer
COEA	Cost and Operational Effectiveness Analysis
COMO	Combat Oriented Maintenance Organization
CPA	Chairman's Program Assessment
CPFF	Cost-Plus-Fixed-Fee
CPIF	Cost-Plus-Incentive-Fee
CPR	Cost Performance Report
CPU	Central Processing Unit
CR	Clarification Request
CRA	Continuing Resolution Authority
CRS	Component Repair Squadron
CRS	Computer Resources Support
CSA	Configuration Status Accounting
CSC	Computer Software Component

CSCI	Computer Software Configuration Item
CSU	Computer Software Unit
CTO	Contracting Training Office
CUT	Cross Utilization Training
C3	Command, Control and Communication
DAB	Defense Acquisition Board
DACM	Defense Acquisition Career Management
DAE	Defense Acquisition Executive
DAL	Data Accession List
DAU	Defense Acquisition University
DBOF	Defense Business Operations Fund
DCAI	Defense Contract Audit Institute
DCMAO	Defense Contract Management Area Operations
DCMC	Defense Contract Management Command
DCPSO	Defense Logistics Agency Civilian Personnel Service Support Office
DDMO	Defense Data Management Office
DDR&E	Director of Defense Research and Engineering
DEA	Data Exchange Agreement
DEM/VAL	Demonstration and Validation
DEPSECDEF	Deputy Secretary of Defense
DFARS	Defense Federal Acquisition Regulation Supplement
DFAS	Defense Finance and Accounting Service
DID	Data Item Description
DLA	Defense Logistics Agency
DMMIS	Depot Maintenance Management Information System
DMO	Data Management Officer
DoD	Department of Defense
DoDD	Department of Defense Directive
DoDI	Department of Defense Instruction
DoDM	Department of Defense Manual
DoN	Department of the Navy
DPG	Defense Planning Guidance
DPML	Deputy Program Manager for Logistics
DPRB	Defense Planning and Resources Board
DPRO	Defense Plant Representative Office
DR	Deficiency Report
DRRB	Data Requirements Review Board
DS	Directorate of Distribution
DSMC	Defense Systems Management College
DSSP	Defense Standardization & Specifications Program
DT&E	Development Test and Evaluation

DTS   Defense Transportation System

EAC   Estimate at Completion
ECP   Engineering Change Proposal
ECR   Embedded Computer Resources
EDMO   Engineering Data Management Officer
EEPROM   Electronically Erasable Programmable Read Only Memory
EMD   Engineering and Manufacturing Development (or EM&D)
EMS   Equipment Maintenance Squadron
EPA   Economic Price Adjustment
EPA   Environmental Protection Agency
ERM   Environmental Resource Management
ESC   Electronic Systems Center
ESEP   Engineering and Science Exchange Programs
EUCOM   European Command

FAR   Federal Acquisition Regulation
FCA   Functional Configuration Audit
FFP   Firm Fixed-Price
FMS   Field Maintenance Squadron
FMS   Foreign Military Sales
FMSO   Foreign Military Sales Order
FORTRAN   FORmula TRANslation (Computer Language)
FOT&E   Follow-On Operational Test and Evaluation
FOTRS   Follow-On Tactical Reconnaisance System
FPI   Fixed-Price Incentive
FQI   Federal Quality Institute
FQR   Formal Qualification Review
FSD   Full Scale Development
FY   Fiscal Year
FYDP   Future Years Defense Program

GAO   General Accounting Office
GBL   Government Bill of Lading
GBU   Guided Bomb Unit
GFP   Government Furnished Property
GM   General Manager (Schedule)
GOCO   Government Owned Contractor Operated
GS   General Schedule
GSBCA   General Services Board of Contract Appeals

HAC   House Appropriations Committee

HASC	House Armed Services Committee
HOL	Higher Order Language
HSC	Human Systems Center
HWCI	Hardware Configuration Item
IAW	In Accordance With
ICA	Independent Cost Analysis
ICAF	Industrial College of the Armed Forces
ICASE	Integrated Computer-Aided Software Engineering
ICBM	Intercontinental Ballistic Missile
ICD	Interface Control Document
ICS	Independent Cost Study
ICWG	Interface Control Working Group
IEP	Information Exchange Program
IFB	Invitation for Bids
IG	Inspector General
ILS	Integrated Logistics Support
ILSM	Integrated Logistics Support Manager
ILSP	Integrated Logistics Support Plan
IM	Item Manager
IMIP	Industrial Modernization Incentive Program
IOC	Initial Operational Capability
IOT&E	Initial Operational Test and Evaluation
IP	Issue Paper
IPA	Integrated Program Assessment
IPS	Integrated Program Summary
IRMC	Information Resources Management College
IRS	Interface Requirements Specification
ISA	Independent Schedule Assessment
ISR	Independent Sufficiency Review
IWSM	Integrated Weapon System Management
IWSMP	Integrated Weapon Systems Management Plan
JCS	Joint Chiefs of Staff
JDAM	Joint Direct Attack Munition
JMA	Joint Mission Analysis
JOPS	Joint Operation Planning System
JPATS	Joint Primary Aircraft Training System
JROC	Joint Requirements Oversight Council
JSOW	Joint Standoff Weapon
JSTARS	Joint Surveilance and Target Attack Radar System
JSTPS	Joint Strategic Target Planning System

LCC	Life-Cycle Cost
LFT&E	Live Fire Test and Evaluation
LOA	Letter of Offer and Acceptance
LRIP	Low-Rate Initial Production
LSA	Logistics Support Analysis
LSAR	Logistics Support Analysis Record
LTC	Lowry Training Center
MAC	Military Airlift Command
MAJCOM	Major Command
MASDC	Military Aircraft Storage and Disposition Center
MBI	Major Budget Issue
MCCR	Mission Critical Computer Resources
MGM	Materiel Group Manager
MICAP	Mission Capability
MIL-STD	Military Standard
MMS	Munitions Maintenance Squadron
MNS	Mission Need Statement
MOA	Memorandum of Agreement
MOC	Maintenance Operation Center
MOU	Memorandum of Understanding
NAMTO	Navy Acquisition Management Training Center
NATO	North Atlantic Treaty Organization
NAVSUP	Naval Supply Systems Command
NCA	National Command Authorities
NFCTC	Naval Facilities Contracts Training Center
NMSD	National Military Strategy Document
NPRDC	Navy Personnel Research and Development Center
NSC	National Security Council
NWAC	Naval Warfare Assessment Center
O&M	Operation and Maintenance
OASN(RD&A)	Office of the Assistant Secretary of the Navy for RD&A
OJCS	Office of the Joint Chiefs of Staff
OMB	Office of Management and Budget
OMS	Organizational Maintenance Squadron
OPR	Office of Primary Responsibility
ORD	Operational Requirements Document
OSD	Office of the Secretary of Defense
OT	One-Time

OT&E	Operational Test and Evaluation
PAT&E	Production Acceptance Test and Evaluation
PB	President's Budget
PBD	Program Budget Decision
PCA	Physical Configuration Audit
PCO	Procuring Contracting Officer
PDCA	Plan-Do-Check-Act
PDL	Programming Design Language
PDM	Program Decision Memorandum
PDM	Programmed Depot Maintenance
PDR	Preliminary Design Review
PE	Program Element
PEM	Program Element Monitor
PEO	Program Executive Officer
PFR	Program Financial Review
PGM	Product Group Manager
PHS&T	Packaging, Handling, Storage & Transportation
PM	Program Manager
PMB	Performance Measurement Baseline
PMD	Program Management Directive
PMP	Program Management Plan
PMRT	Program Management Responsibility Transfer
PO	Program Office
POM	Program Objective Memorandum
PPBS	Planning, Programming, and Budgeting System
PPI	Pre-Planned Product Improvement
PR	Purchase Request
PRO	Plant Representative Office
PROM	Programmable Read Only Memory
PRR	Production Readiness Review
PTO	Participating Test Organization
QOT&E	Qualifying Operational Test and Evaluation
QT&E	Qualification Test and Evaluation
R&D	Research and Development
R&M	Reliability and Maintainability
RAM	Random Access Memory
RCM	Requirements Correlation Matrix
RD&A	Research, Development and Acquisition (Navy)
RDT&E	Research, Development, Test and Evaluation

REMCO	Rear Echelon Maintenance Combined Operations
RFP	Request for Proposal
ROM	Read Only Memory
ROM	Rough Order of Magnitude (Cost)
RTO	Responsible Test Organization
S&T	Science and Technology
SAB	Scientific Advisory Board
SAC	Senate Appropriations Committee
SAC	Strategic Air Command
SAE	Service Acquisition Executive
SAF	Secretary of the Air Force
SAR	Selected Acquisition Reports
SASC	Senate Armed Services Committee
SBTW/AA	Space-Based Tactical Warning and Attack Assessment
SCN	Specification Change Notice
SDF	Software Development File
SDI	Strategic Defense Initiative
SDP	Software Development Plan
SE	Support Equipment
SECDEF	Secretary of Defense
SEDS	System Engineering Detailed Schedule
SEMP	Systems Engineering Management Plan
SEMS	System Engineering Master Schedule
SFR	System Functional Review
SFW	Sensor-Fused Weapon
SMSC	Space and Missiles Systems Center
SOW	Statement of Work
SPO	System Program Office
SPD	System Program Director
SRAM	Short Range Attack Missile
SRD	System Requirements Document
SRR	System Requirements Review
SRS	Software Requirements Specification
SSA	Source Selection Authority
SSAC	Source Selection Advisory Council
SSDD	System/Segment Design Document
SSEB	Source Selection Evaluation Board
SSP	Source Selection Plan
SSR	Software Specification Review
START	Strategic Arms Reduction Treaty
STD	Standard

STINFO	Scientific and Technical Information
SVR	System Verification Review
SYDP	Six Year Defense Program
T&E	Test and Evaluation
TAC	Tactical Air Command
TAP	Technology Area Plan
TBD	To Be Determined
TCO	Termination Contracting Officer
TCP	Task Change Proposal
TCTO	Time Compliance Technical Order
TDP	Technical Data Package
TEMP	Test and Evaluation Master Plan
TEO	Technology Executive Officer
TIM	Technical Interchange Meeting
TM	Tech Manuals
TO	Technical Order
TPM	Technical Performance Measurement
TPWG	Test Plan Working Group
TQ	Total Quality
TQM	Total Quality Management
TRR	Test Readiness Review
TSSAM	Tri-Service Standoff Attack Missile
TTP	Technology Transition Plan
UCF	Uniform Contract Format
USD(A)	Under Secretary of Defense for Acquisition
UVPROM	Ultra-Violet Programmable Read Only Memory
WBS	Work Breakdown Structure